Preface

POPULATION GENETICS IS OFTEN THOUGHT TO BE A DIFFICULT SUBJECT. To some extent, difficulties are inevitable in a field where some quite basic points are controversial. However, problems are most acute when theoretical points are discussed, despite the fact that there has been very little controversy over the mathematics. In my experience, the actual mathematical manipulations rarely cause much difficulty. Rather it is that the biologist, lacking the physicist's or chemist's experience in "reading" mathematical formulae, finds it difficult to appreciate what is happening in a mathematical treatment and to grasp the implications of the results obtained, when these are given in mathematical form. Accordingly, I have followed a procedure, which students seem to find helpful, of giving a rough-and-ready verbal treatment of a problem before attempting a much more exact mathematical treatment; when the results of the latter are not readily interpretable, I have given an elucidation. Another problem which often concerns students is the reliability of results obtained using approximate methods; I have, therefore, discussed this in fair detail in critical cases.

When dealing with controversial issues, I have done my very best to be fair. To conceal one's opinions entirely would probably make for a very dull book. I trust, however, that I have given enough for the reader previously unfamiliar with these controversies to form his own judgement.

To acknowledge all those who have so greatly assisted my understanding of population genetics would mean a very lengthy list. I should, however, particularly mention E. J. Machin, who introduced me to the subject in my schooldays, and the inspiring lectures of Dr. A. R. G. Owen. To express my debt to Sir Ronald Fisher would require literary powers far beyond my own; in his presence, indeed, "meadow, grove, and stream,/The earth, and every common sight/To me did seem/Apparelled in celestial light".

v

No author could be blessed with more helpful colleagues. I should particularly like to thank Dr. P. D. S. Caligari for his willingness to give unlimited time to discussing problems and their presentation, and for his many valuable suggestions. I am also indebted for encouragement, assistance and advice to Prof. J. L. Jinks, Rev. Dr. L. J. Eaves, Dr. M. J. Kearsey, Dr. A. J. Birley, Dr. G. H. Jones, Dr. A. J. Cornish-Bowden, Dr. N. Goodchild, Mr. I. J. Mackay and Mr. J. P. Gibson. Prof. Bryan Clarke was kind enough to read and comment on the section on *Cepaea* (although he would not necessarily agree with all that appears there). Finally, I should like to record my debt to my students. By their comments, queries, criticisms and (very occasional!) errors, they have guided me to what I believe to be a greatly improved understanding of the subject and its presentation. Of course, I take sole responsibility for anything in the book which is incompetent, irrelevant or immaterial.

J.S.G.

Population Genetics

TERTIARY LEVEL BIOLOGY

A series covering selected areas of biology at advanced
undergraduate level. While designed specifically for course
options at this level within Universities and
Polytechnics, the series will be of great value to
specialists and research workers in other fields who require
a knowledge of the essentials of a subject.

Titles in the series:

Biological Membranes	Harrison and Lunt
Water and Plants	Meidner and Sheriff
Comparative Immunobiology	Manning and Turner
Methods in Experimental Biology	Ralph
Experimentation in Biology	Ridgman
Visceral Muscle	Huddart and Hunt
An Introduction to Biological Rhythms	Saunders
Biology of Nematodes	Croll and Matthews
Biology of Ageing	Lamb
Biology of Reproduction	Hogarth
An Introduction to Marine Science	Meadows and Campbell
Biology of Fresh Waters	Maitland
An Introduction to Developmental Biology	Ede
Neurosecretion	Maddrell and Nordmann
Physiology of Parasites	Chappell
Biology of Communication	Lewis and Gower

Population Genetics

J. S. GALE

Lecturer in Genetics
University of Birmingham
England

Blackie
Glasgow and London

Blackie & Son Limited
Bishopbriggs
Glasgow G64 2NZ

Furnival House
14–18 High Holborn
London WC1V 6BX

International Standard Book Numbers
Hardback 0 216 91000 5
Paperback 0 216 91001 3

Filmset by Advanced Filmsetters (Glasgow) Limited
Printed in Great Britain by
Thomson Litho Ltd., East Kilbride, Scotland

Contents

vii

In Memoriam—E. J. Machin
Best of Schoolmasters

CHAPTER ONE

INTRODUCTION

But pardon, gentles all,
The flat unraised spirit, that hath dared,
On this unworthy scaffold to bring forth
So great an object...
...For the which supply,
Admit me chorus to this history;
Who, prologue-like, your humble patience pray,
Gently to hear, kindly to judge, our play.

William Shakespeare, *King Henry V*

Natural selection

Hermann Bondi once wrote: "There can be no greater merit in a scientific discovery than that before long it should appear odd that it ever was considered a discovery... only those things that have so deeply affected our thinking and so thoroughly changed our outlook that we cannot think without them have really entered the spirit of the human race". These remarks, although made in another context, are entirely appropriate to the theory of evolution by natural selection. This theory is accepted (provisionally, as with any other scientific theory) because it conforms with the rest of our biological knowledge, because it explains the universal adaptation of living organisms to their environments, because specific examples of evolution by natural selection have been demonstrated in practice and, finally, because the theory can be formulated in a sufficiently precise way to be tested in practice. Thus the theory states that the process of natural selection is sufficient to bring about the changes that have taken place during the evolution of a particular character *in the time actually available*. Hence for a critical test of the notion that the evolution of some specific character has come

1

about through the agency of natural selection, we have first to show that the character is under genetic control, next to determine the selective advantage of the character to those individuals possessing it and, finally, to demonstrate that the magnitude of this advantage is large enough to have led to the changes that have actually occurred over a defined period of time. Critical tests of this kind can be carried out in cases of present-day evolution (although the practical difficulties are often formidable); thus the theory has the attribute "falsifiability in principle" generally held to be an essential feature of any useful scientific theory.

Problems studied by population geneticists

We shall centre all our later discussion on a number of problems critical to our understanding of the evolutionary process. Among these problems will be: Does natural selection within a group of organisms living in a particular habitat usually lead to uniformity or to diversity? What is the relationship between magnitude of selective advantages and rate of evolutionary change? Can natural selection give rise to self-sacrificing behaviour? How widespread are traits that convey some selective advantage?

Thus some workers, while accepting that obviously adaptive characters have spread through the agency of natural selection, consider that many other characters have spread purely by chance, it being a matter of indifference to the organism whether or not it carries the genes giving rise to such characters. This is the celebrated *neutral gene* theory. The opposing view is that natural selection is of near-universal occurrence and must be invoked in order to explain almost all the changes that have taken place in evolution. As we shall see, the neutralist and selectionist viewpoints lead to quite different predictions about natural populations, and many active attempts are being made to reach a decisive solution to this problem. It is quite wrong to suppose, as is occasionally done by those not familiar with the actual practice of population genetics, that any result not immediately explicable by natural selection is simply written off as a case where unknown forms of natural selection have operated. Readers of *Gulliver's Travels* will recall that such an appeal to occult causes was beneath the notice even of the imbecile scholars of the King of Brobdingnag. Selectionists, in common with other researchers, wish not to dismiss problems but to solve them; that is, to study problems, formulate hypotheses from which predictions can be made, and find out whether these predictions hold in practice.

Approaches to our problems

It might be supposed that the fossil record, so important for the elucidation of many aspects of evolution, would greatly assist the solution of the problems we have formulated. Unfortunately, much information essential to us is unavailable in the fossil record, which is therefore of limited help. Thus we do not know the magnitude of the selective advantage of successful types as compared with the unsuccessful types they replaced. We do not know the size of past interbreeding populations or the amount of inbreeding. However, we shall occasionally find the fossil record helpful; most useful in our context will be the study of rates of evolution. The fossil record suggests that evolution was a comparatively slow process in many cases. As we shall see, this observation in itself is compatible with either the neutralist or the selectionist theories, but an attempt to discriminate between these theories has been made by calculating for a given protein (e.g. haemoglobin) the average number of amino acid substitutions per amino acid site per year for a given line of descent (often called the rate of evolution of the protein for that line of descent). We shall discuss the details of the argument in chapter 6, but the basic idea is that if evolution of the protein was brought about by natural selection, the rate of evolution would vary from one line of descent to another. If, however, most of the amino acid substitutions were of no adaptive significance, it can be shown (given certain assumptions) that the rate would be about the same for different lines.

A much more useful approach, indicated briefly earlier, is the study of present-day natural populations. Here we can, at least in principle, estimate the magnitude of any factor (for example, population size, intensity of natural selection) relevant to the evolution of the population under investigation. However, such study of natural populations ("ecological genetics") is not always easy. One difficulty results from the gross heterogeneity of the natural habitat, a phenomenon which the reader can easily verify for himself. Frost cover, for example, may vary markedly over short distances. Thus natural selection may act quite differently in adjacent portions of the habitat, a genotype favoured in one portion being at a disadvantage in another. Similarly, the most successful genotype in a particular year might fail in another year when, say, climatic conditions were different.

In an effort to avoid these complications, experimenters often set up population cages in the laboratory, with a view to studying evolution under more or less controlled conditions. The advantages of this approach are fairly obvious: different groups of individuals taken from a specific (natural or cage) population can be subjected to a variety of defined environmental variables, such as specified temperatures, humidity or food, and these treatments can be replicated so that the effects of other factors such as chance or uncontrolled environmental changes can be estimated. The main difficulty with cage experiments is that simplifying the natural habitat is, of course, falsifying it; for any individual experiment, we must try to decide whether the results would hold in nature. Despite this difficulty, cage experiments can be very helpful. To take a famous example (see chapter 10), suppose we observe that the frequencies of specific genotypes show marked fluctuations in the course of the year, the same fluctuations being found for several years in succession. We hypothesize that the relative selective advantages or disadvantages of the different genotypes fluctuate during the year as a result of changes in environmental factors such as food supply; we attempt, from such knowledge as we have of the biology of the organism concerned, to list such environmental factors. If we have correctly identified a relevant environmental factor, it should be possible to mimic the fluctuations in genotype frequency which are found in nature by varying this factor in cage experiments, an approach that has met with considerable success.

However, neither natural nor cage populations can in practice supply answers to all relevant problems. To take an example: it is often asked "can a selective advantage be so small that it does not matter?" More precisely, we might guess that if the selective advantage accruing to an individual from the possession of a particular allele is small, the allele will be "effectively neutral", that is, it will change in frequency in virtually the same way as does a neutral allele. We ask whether our guess is correct and if so how small must be the selective advantage for the allele to be effectively neutral. In principle, the question could be answered experimentally, by studying changes in frequency of alleles conveying small selective advantages. In practice, such an experiment is virtually impossible to carry out, since to estimate the selective advantage with the required degree of precision would involve an experiment on a truly enormous scale.

Thus there are points of great importance to us which cannot be investigated by observation or experiment. Hence we are forced to another approach—mathematical consideration of relatively simple models of evolution in action. Beginners often dislike such theoretical studies. Insofar as this is not due merely to a distaste for mathematics as such (a distaste which we shall try to help such readers to overcome) the main objection appears to be as follows. Natural situations may well be very complicated, but the theory has not advanced sufficiently far to take account of most of these complications; the models used are gross over-simplifications of reality and so are liable to give very misleading conclusions. There is some justice in this objection. Obviously the answers we get will depend on the assumptions made, and these may sometimes be arbitrary or artificial.

Nevertheless, the objection is easily over-stated. Some results (e.g. that in large populations neutral alleles change in frequency very slowly) are almost certainly proof against any adjustments that might be made in the model. In the case of others, we can try adjusting the model in various ways and see what happens. We suggest that the reader approach the theory with a critical but open mind; ultimately, he must decide for himself the relevance of the results we shall give—theoretical, observational or experimental. It may help, however, at this preliminary stage, to recall an analogous situation in the physical sciences. When first attempting to study the flight of a projectile, we often ignore in the first instance the obviously relevant factor of air resistance; later, we may take account of the latter but ignore the rotation of the earth, although this is important in long-range gunnery. Yet the over-simplified models give useful results.

In summary, four main approaches are available: investigation of natural populations, of cage populations, of the fossil record and of theoretical models. We shall concentrate our discussions on specific problems, and for any such problem use any approach or approaches which seem appropriate for its resolution. For an alternative standpoint, taken by some authors, in which a particular approach is built up into a coherent body of knowledge with only occasional references to other approaches, the reader is referred to the literature (some suggestions for further reading are given at the end of this book).

The Mendelian population

In the interests of clarity, we define a Mendelian population as a group of inter-mating individuals. Thus an isolated group of 100 *strictly self-pollinating* plants living in the same area, although a population in the ecologist's sense, is not a single Mendelian population, size 100, but 100 Mendelian populations each of size one, since individuals never exchange genes with other individuals. On the other hand, two groups of outbreeding plants which are geographically separated but which sometimes inter-mate by the passage of pollen from one group to the other, are all members of just one Mendelian population. However, for brevity's sake we shall (as hitherto) use the term *population* as short for Mendelian population, except in cases where confusion might arise.

We shall be concentrating subsequently on factors affecting the genetic composition of such a Mendelian population. Unless stated otherwise, we shall take it (for simplicity) that this population has constant size N and that different generations do not overlap; usually, individuals will be diploid. We shall often consider just one locus at a time, with just two alleles at a locus. If at some time there are NP individuals of genotype $\underline{AA}$, $2NQ$ individuals of genotype $\underline{Aa}$ and NR individuals of genotype $\underline{aa}$ (the use of $2NQ$ rather than NQ for $\underline{Aa}$ will simplify later calculations) then the proportions of these three genotypes (the "genotype frequencies") at that time are

Genotype	$\underline{AA}$	$\underline{Aa}$	$\underline{aa}$	Total
Frequency	P	$2Q$	R	1

the total being unity since we have assumed only two alleles present in the population. These genotype frequencies summarize the genetic composition of the population at the particular time. Since the frequencies add to unity, it would be sufficient to give two of them, the remaining frequency being immediately obtainable by subtraction.

A simpler summary is given by the frequencies of the two alleles (the *allele frequencies*—often referred to, slightly loosely, as the "gene frequencies"). Since individuals are diploid, there will be $2N$ alleles at the locus in the population as a whole; of these alleles $(2NP+2NQ)$ will be $\underline{A}$, $(2NQ+2NR)$ will be $\underline{a}$, so that the allele frequencies are

Allele	$\underline{A}$	$\underline{a}$	Total
Frequency	$P+Q = p$ (say)	$Q+R = q$ (say)	1

Since $p+q = 1$, it is sufficient to give either p or q. Whenever possible, we shall give the simpler summary in terms of one allele frequency (usually p) rather than the more complicated summary in terms of two genotype frequencies.

Factors affecting the genetic composition of populations

The evolutionary process is often envisaged in terms of changes in allele frequencies or in genotype frequencies; the multiplicity of phenomena bringing about changes in genotype frequency are grouped under the headings: mating system, chance, natural selection, mutation. If, as often happens, the population is divided into sub-populations, with members of any sub-population breeding freely *inter se* but in a more restricted way with members of other sub-populations, then different sub-populations may differ from one another in their genetic composition. In such a case, an additional phenomenon, the migration of individuals or gametes (e.g. pollen) from one sub-population to another, can have an important effect on frequencies.

It is easiest, in the first instance, to consider each group of phenomena on its own, ignoring for the moment the effects of the others.

A fundamental consequence of Mendelian inheritance may be stated informally as: the hereditary mechanism, of itself, does not change the allelic composition of a population. More formally we say: in the absence of chance effects, natural selection and mutation, the allele frequencies remain constant over generations, whatever the mating system; for, with no chance effects, selection or mutation, genotype frequencies and hence allele frequencies p, q remain unchanged throughout the life cycle. Now (still with the same preconditions) consider gametogenesis: the proportion of gametes containing $\underline{A}$ will be p, the proportion of gametes containing $\underline{a}$ will be q, that is, gametic frequencies equal allele frequencies. Irrespective of the way in which these gametes are combined into zygotes giving rise to the next generation, allele frequencies in this next generation will remain unchanged at values p, q.

On the other hand, the mating system will determine how gametes are combined into zygotes, and will thus affect zygotic frequencies, as we shall show shortly by consideration of two different mating systems. Some examples of different mating systems are: random mating (the chance that an individual mates with a particular genotype equals the frequency of that genotype in the population); positive assortative mating (like phenotypes tend to mate); inbreeding (close relatives tend to mate)—we consider this more fully in chapter 2. Note that the mating system is not necessarily the same for different loci. For example, when choosing a mate, a human being may note many aspects of the potential partner's phenotype, but hardly their blood group. Not surprisingly, then, we find in our species positive assortative mating for stature, whereas blood groups provide excellent examples of random mating.

Under random mating, genotype frequencies among zygotes are very simply related to the frequencies of the gametes from which these zygotes were derived. Suppose gametic frequencies are $\underline{A}$ p, $\underline{a}$ q. Now random mating implies random union of gametes (for a discussion of this point, see Edwards 1977). Thus in the absence of mutation, selection on the gametes or chance effects, we have

Type of union		Frequency	Derived zygote
Egg	Sperm		
$\underline{A}$	$\underline{A}$	p^2	$\underline{AA}$
$\underline{A}$	$\underline{a}$	pq	$\underline{Aa}$
$\underline{a}$	$\underline{A}$	qp	$\underline{Aa}$
$\underline{a}$	$\underline{a}$	q^2	$\underline{aa}$

so that frequencies of the zygotes are

$$\underline{AA} \quad p^2, \quad \underline{Aa} \quad 2pq, \quad \underline{aa} \quad q^2$$

The new allele frequencies are

$$\underline{A} \quad p^2 + pq = p(p+q) = p$$
$$\underline{a} \quad q^2 + pq = q(q+p) = q$$

since $p + q = 1$. Thus the new allele frequencies are equal to the gametic frequencies, and hence to the allele frequencies one generation preceding, as we have already established earlier for the more general case.

With the allele frequencies unchanged, the new gametic frequencies will be the same as those one generation preceding. It is apparent, then, that genotype frequencies, established in one generation at p^2, $2pq$, q^2 remain at these values indefinitely. This result, known as the Hardy–Weinberg law, may be stated in the form

"a sufficient condition for no evolution to occur within a Mendelian population is that mutation, selection and chance effects are all absent and that mating is at random".

If we wish to apply the law to a population in the ecologist's sense, we must have the additional precondition "no migration".

The Hardy–Weinberg law is best regarded as a special case of the fundamental principle, given earlier, that the hereditary mechanism, of itself, does not change allele frequencies. The constancy of the genotypic frequencies (given no mutation, chance effects or selection) then follows from the presence of random mating.

With other mating systems, genotype frequencies can change quite markedly from one generation to the next, although this does not always happen. Consider, for example, a flowering plant reproducing partly by selfing, partly by outcrossing. Suppose a proportion s of the seed arises from selfing, the remaining proportion $(1-s)$ arising from random mating. Once again we suppose mutation, selection and chance effects negligible. Write the genotype frequencies in some generation (generation t, say) as

$$\begin{array}{ccc} \underline{AA} & \underline{Aa} & \underline{aa} \\ P & 2Q & R \end{array}$$

On selfing, $\underline{AA}$ gives $\underline{AA}$ offspring only and $\underline{aa}$ gives $\underline{aa}$ offspring only, whereas $\underline{Aa}$ gives offspring $\frac{1}{4}\underline{AA}$, $\frac{1}{2}\underline{Aa}$, $\frac{1}{4}\underline{aa}$; hence in the next generation, generation $(t+1)$, the frequency of $\underline{Aa}$ arising from selfing is

$$s \times 2Q \times \tfrac{1}{2} = sQ$$

The frequency of $\underline{Aa}$ arising from random mating is

$$(1-s)2pq$$

In all, then, the frequency of $\underline{Aa}$ in generation $(t+1)$ is

$$sQ + (1-s)2pq = 2Q', \text{ say}$$

the change in frequency being

$$\begin{aligned} 2Q' - 2Q &= (s-2)Q + (1-s)2pq \\ &= (s-2)Q + (2-s)Q_e \\ &= (1-\tfrac{1}{2}s)(2Q_e - 2Q) \end{aligned}$$

where Q_e is defined as

$$Q_e = \frac{(1-s)2pq}{2-s}$$

It is apparent then that if in generation t the frequency of $\underline{Aa}$, namely $2Q$, happened to equal $2Q_e$, the change in frequency would be zero; the frequency of $\underline{Aa}$ would remain constant over successive generations. Otherwise, the frequency of $\underline{Aa}$ changes. To see the direction of change, note first that s, the proportion of selfing, is positive but less than unity, so that $(1-\tfrac{1}{2}s)$ is necessarily positive in sign. Hence, if $2Q$ is *less* than

$2Q_e$, so that $(2Q_e - 2Q)$ is positive, the change in frequency is positive, that is, the frequency of A̲a̲ *rises*. Since $(1 - \frac{1}{2}s)$ is less than unity, the rise is less than $(2Q_e - 2Q)$. Hence in successive generations, the frequency of A̲a̲ gradually approaches (but never exceeds) $2Q_e$. Similarly, if in generation t, $2Q$ is *greater* than $2Q_e$, the frequency of A̲a̲ *falls* over successive generations to $2Q_e$.

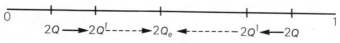

Thus ultimately the frequency of A̲a̲ (and also of A̲A̲, a̲a̲) settles down at a constant value, the heterozygote frequency being less than that obtaining under random mating by a factor

$$\frac{2(1-s)}{2-s} \quad \text{(Haldane 1924a)}$$

Common sense, of course, indicates that the heterozygote frequency is lower than under random mating, but (usually) rather exaggerates the degree of reduction. For example, in the long-headed poppy, *Papaver dubium*, s is about 0.75 (Humphreys and Gale 1974). Substitution of this value in the formula above gives a value 0.4 for the factor. Thus, although outcrossing is only 25%, the heterozygote frequency is 40% of that which would be found under random mating. We stress this point as exemplifying the need for mathematical treatment of problems in population genetics. Intuitive verbal discussions are helpful in giving a feel for a problem, but may give rather distorted conclusions (and occasionally are downright misleading). The reader is advised to repeat our calculations for a few other values of s.

We turn now to phenomena which affect allele as well as genotype frequencies: chance, mutation, selection, migration. The evolutionist will naturally be interested in the long-term consequences of these but since, as Lord Keynes put it, "in the long run we are all dead", the rate of change in allele frequency is also critical. Mutation, for example, generates new alleles at a very low rate; the proportion of gametes containing a newly arisen allele at a specified locus being (probably) in the region 1 in 100 000 to 1 in a million for eukaryotic organisms. An unknown proportion of such alleles, perhaps a large majority, are sufficiently disadvantageous to their possessors to be eventually eliminated from the population by natural selection. Thus, even if for the moment we ignore the effects of chance, it is apparent that allele frequencies will change only very slowly indeed as a result of mutation alone. In fact, as we shall see in chapter 3, the bulk of new neutral mutations are lost from the population purely by chance; for disadvantageous mutants, selection will reinforce this effect.

Changes in allele frequencies brought about by chance are known as random genetic drift. Suppose allele frequencies among the zygotes in some generation are $\underline{A}\,p, \underline{a}\,q$. As we have seen, in the absence of chance, mutation or selection, the frequencies of the gametes which actually give rise to the next generation would also be p, q. By chance, however, these frequencies, which are also the allele frequencies among the zygotes in the next generation, will usually differ a little from p, q. This process is analogous to more familiar examples of *random sampling*. If the Mendelian population consists of N individuals, so that at a given locus there are $2N$ alleles in all, our process may be conceived as the drawing of a random sample of alleles, size $2N$, from a conceptual population, indefinitely large in size, of possible alleles; in this conceptual population, allele frequencies are $\underline{A}\,p, \underline{a}\,q$. From the usual properties of random samples, it is apparent that the smaller the size of the sample of alleles, the more likely are the allele frequencies to depart from those in the conceptual population. In plain English, chance changes of allele frequency in passing from one generation to the next will, in general, be more marked in small populations than in large. In the latter case, changes will normally be very slight.

It is important to realize that "nature has no memory"; thus, if by chance the frequency of $\underline{A}$ falls in one generation, the population preserves no memory of this fall. Every generation starts anew, irrespective of how it arrived at its present composition, and there is no tendency for a fall in one generation to lead to a compensating rise in the next although, of course, such a rise may occur by chance. Thus, over a number of generations, changes in frequency may, by chance, be predominantly in one direction. If the population is very small, so that the change in a single generation will often be large, quite marked changes in allele frequency over a number of generations may arise in this manner. In large populations, very little change is likely over any but a very large number of generations, as we shall demonstrate in a more rigorous manner in chapter 3.

In contrast to the preceding, natural selection may bring about very rapid changes in allele frequency in any population if the intensity of selection is sufficiently great. To take an obvious but revealing example, if we had a population mostly $\underline{aa}$, $\underline{Aa}$ with only a few $\underline{AA}$, and the environment changed in such a way that only AA individuals could survive, the frequency of the A allele would rise at once from its previous low value to value unity. Although with mild selection changes will be much slower, a practical consequence is that if we observe a rapid change in allele frequency in any but a very small Mendelian population, we can at once attribute this result to natural selection. We return to this topic in chapter 8, when we shall be particularly interested in the speed of change under rather mild selection, a topic which was very prominent in the early stages of the development of population genetics.

In the example just given, the population ended up all AA, that is, genetically uniform (*monomorphic*) at the locus concerned. If, however, AA was much the fittest genotype in some years, whereas aa was much the fittest in other years, with Aa roughly intermediate in all years, the outcome is less certain—we might imagine that the population would remain genetically variable (*polymorphic*) with all three genotypes present for an indefinitely long period of time. In fact (chapter 10), polymorphism will obtain under quite a variety of selective regimes. As a result of work carried out in the 1960s, it became apparent that polymorphism is a very common phenomenon. To what extent this polymorphism is due to natural selection and to what extent to the chance gradual spread of neutral genes, is an unsolved problem which excites much passion, but which must ultimately be solved by careful collection of appropriate data.

Summary

The domain of population genetics is the genetic composition of natural populations, past and present. This composition is summarized in the frequencies of alleles and genotypes. Allele frequencies change very slowly under the action of mutation and chance. Natural selection can bring about rapid change in allele frequency. Depending on the type of selection operating, natural selection may lead to genetic uniformity or to diversity. Knowledge of the mating system is critical for an understanding of the manner in which alleles are combined into genotypes.

Research in population genetics is directed towards problems whose resolution is essential for a just appreciation of the evolutionary process. Prominent among these are the relative importance of natural selection and chance, the rate of evolutionary change, long-term effects of natural selection. Appeals to present-day populations, natural and laboratory, to theoretical analysis and (to a limited extent) to the fossil record provide the relevant evidence.

CHAPTER TWO

SELFING

If undesirable characters are shown after inbreeding, it is only because they already existed in the stock...inbreeding is no more to be blamed than the detective who unearths a crime.

East and Jones, *Inbreeding and Outbreeding* (1919)

Inbreeding

Casual observations on the effects of inbreeding (that is, the mating of close relatives) were probably first made a very long time ago by early breeders of domestic animals. However, Mendel (1865) was the first to describe these effects in a quantitative manner. Consider selfing. A homozygote on selfing gives homozygous offspring only, whereas only half of the offspring of a selfed heterozygote (on average) will be heterozygous. Suppose we take a set of individuals, some of which are homozygous and some heterozygous at a given locus, and self every individual. We expect a reduction in the proportion of heterozygotes as we pass from one generation to the next. If we continue to self successive generations, we expect the proportion of heterozygotes to fall steadily and eventually to reach zero. Mendel gave a formula from which the "expected" proportion of heterozygotes (in a sense we shall define precisely shortly) in any such generation could be predicted. Now this reduction in the proportion of heterozygotes should obtain at every locus; we expect that eventually all individuals will be homozygous at every locus (apart from the occasional heterozygote resulting from new mutation).

We shall devote most of this chapter to discussing the prospects for such universal homozygosity, given a specified number of successive selfings. Of course, the term "inbreeding" covers a wide range of systems of mating, for example, brother-sister mating, parent-offspring mating. Some of these other systems were investigated theoretically in considerable detail quite early in the history of genetics, mainly by Pearl, Jennings and Robbins, writing over the period 1914–1918. Since that

14

time, theoretical studies of the genetical effects of almost every conceivable system have been made by many workers, of whom Wright, Haldane, Malécot and Fisher are perhaps the most prominent. However, since an individual is more closely related to himself than to any other individual, we expect that the proportion of heterozygotes will fall much more rapidly under selfing than under any other system, and this turns out to be so. We shall therefore discuss in detail selfing only, since this will demonstrate the effects of inbreeding in a particularly dramatic form, with the added advantage that selfing turns out to be much simpler to discuss than any of the other systems.

Inbreeding depression

So far we have tacitly assumed that all genotypes at a given locus are equally viable and fertile (a point fully realized by Mendel). However, this assumption cannot, in general, be justified. In any population, harmful mutations will arise from time to time. A majority of these mutations will be recessive (for reasons we discuss in chapter 9), that is, their harmful effects will not be apparent (or only very slightly apparent) in heterozygotes. Many of these recessive mutants or their descendants are rapidly lost from the population by chance, but in some cases descendants persist for a while before (nearly always) being finally lost by the combined action of chance and selection against the mutant homozygote. In outbred populations, heterozygotes for the mutant will greatly outnumber mutant homozygotes; thus recessive mutants will be "protected" from natural selection to a considerable extent, and will thus persist for much longer than in more inbred populations. Occasionally, the frequency of heterozygotes for a harmful recessive allele can, owing to chance, become surprisingly large. This occurs, for example, when an individual heterozygous for a harmful recessive allele happens, for reasons unconnected with his genotype at the locus in question, to leave many descendants. This has been demonstrated very convincingly in our own species.

Some interesting examples are found among a religious sect, the Old Order Amish (McKusick, Hostetler, Egeland and Eldridge 1964; McKusick 1974). In Lancaster County, Pennsylvania, about 13% of members of the sect are heterozygous for the (normally very rare) allele which in homozygous condition gives rise to the very harmful Ellis-van Creveld syndrome. All known heterozygotes were found to descend from a single couple who immigrated in the 1740s and left many descendants.

Of course, such high frequencies of heterozygotes for an apparently unconditionally harmful allele are unusual. In general, harmful recessive alleles at any given locus will be present, if at all, at low frequency determined by the interplay of mutation to the alleles, selection against them when in homozygous form, and random genetic drift (the last factor often leading to extinction of the alleles for most of the time in smallish populations). Nevertheless in practice the cumulative effect over loci can be quite striking; individuals completely free of harmful recessives at every locus seem, in outbreeding species, to be very unusual.

The most comprehensive studies on this point have been carried out in species of *Drosophila* where special breeding techniques make it possible to determine for any given chromosome, such as chromosome II, the proportion of individual chromosomes in the population carrying a harmful recessive. These results are summarized and discussed by Dobzhansky (1970). To make any progress, we must of course make clear what is meant by "harmful". We first define the viability of a genotype as the proportion of individuals of that genotype which survive from zygote to adult. Genotypes conferring zero viability are referred to as lethal, and those with viability between zero and 0.5 as semilethal. Less drastic reductions in viability are a little more difficult to define.

Suppose we make up a reference set of genotypes by random combination of chromosomes randomly chosen from the population under study. The viabilities of these genotypes are then determined; the mean $\bar{x}$ and standard deviation s of these viabilities are then calculated. Any viability lying within the range

$$\bar{x} - 2s, \quad \bar{x} + 2s$$

is said to lie within the normal range (Wallace and Madden 1953). Genotypes with viability below the normal range but above 0.5 are classified as subvital. Thus any genotype in any part of the experiment can be classified as lethal, semilethal, subvital, normal, or even supervital (above the normal range) using the definitions established by the performance of the genotypes in the reference set. By an appropriate breeding programme, it is possible to obtain zygotes homozygous for individual chromosomes in the population under study, thus making any recessive alleles on such chromosomes homozygous. The viabilities of these homozygotes can then be determined and classified. Among much data cited by Dobzhansky, consider as example the results of Dobzhansky, Spassky and Sankaranarayanan on a Californian population of *Drosophila pseudoobscura*. Among the individual chromosome

IIs made homozygous, 33.0% were lethal or semilethal (mostly lethal), 62.6% subvital and only 4.3% normal (less than 0.1% were supervital). Moreover, among the subvital and normal group taken together, 10.6% were female sterile and 8.3% male sterile. Roughly similar results were obtained for chromosomes III and IV, with however a larger proportion of normals (16.3% and 22.3% respectively). Dobzhansky concludes that "very few individuals in nature carry only normal and supervital chromosomes". Although studies of this kind have been most extensive on *Drosophila*, investigations on other genera make it clear that harmful recessives are very widely distributed in natural populations.

This being so, some of the homozygotes produced on inbreeding will have zero or poor fitness. This is the major explanation for *inbreeding depression*, that is the decline of fitness on inbreeding. As indicated earlier, this phenomenon seems to have been recognized for a very long time. Among early scientific studies, the very extensive experiments on flowering plants carried out by Darwin (1876) take pride of place. He demonstrated, for the first time quantitatively, that the offspring of crosses were generally superior to the offspring of selfs in respect to height, weight, fertility and seed output. He fully appreciated that these differences must be related in some way to the hereditary constitution of the individuals concerned; he stated "there is the clearest evidence...that the advantage of a cross depends wholly on the plants differing somewhat in constitution". Of course, Darwin was not acquainted with Mendel's work, so that the explanation of his results came very much later than the original work. Generally, it is clear that inbreeding depression is a very widespread phenomenon, embracing many aspects of fitness. Inbred mice, for example, as compared with outbreds are less viable, bear fewer young, lactate less well, mature later, have a longer gestation period and a shorter breeding life. The reader is referred to Wright (1977) for a detailed summary and discussion of the many experiments demonstrating inbreeding depression. However, "age cannot wither nor custom stale"—the reader is also strongly encouraged to consult Darwin's book.

Difficulties in allowing for natural selection during inbreeding

Now inbreeding depression poses a serious difficulty for any theoretical treatment of the inbreeding process. Consider selfing. If one of the possible homozygotes at a locus is lethal, semilethal or subvital, or has poor fertility, the approach to homozygosity is slowed down, not just at the locus concerned but also at loci linked to it. With other systems of inbreeding, lethality of one homozygote will actually speed up the approach to homozygosity, since the mating $AA \times aa$, which yields heterozygotes only, is prevented. In either case, allowance should clearly be made for the effects of such natural selection. Further allowance must be made for loci at which the heterozygote is superior in fitness to *both* homozygotes (unfortunately there is no general agreement on how often such loci will be encountered—but see chapter 10). Although some progress has been made in dealing with the effects of natural selection during inbreeding (Haldane 1936, 1956; Hayman and Mather 1953, 1956; Reeve 1955, 1957—for a helpful summary see Falconer 1964), the principal difficulty lies in our very inadequate knowledge of selective effects at the loci concerned, even in *Drosophila*, where there are probably many other selective effects apart from those actually detected (Dobzhansky points out, for example, that homozygotes often develop very slowly; such slow developers would often be omitted from the next generation and other unguessed effects may be similarly important). One further complication (as will be apparent later) would be a reduction in chiasma frequency during the inbreeding process, as found in inbred rye (Rees and Thompson 1956). While investigations of long-inbred organisms do indicate that homozygosity is eventually reached in many cases, such as brother-sister mated mice (Billingham, Brent, Medawar and Sparrow 1954; Deol, Grüneberg, Searle and Truslove 1960) a *quantitatively accurate* theory of the approach to homozygosity in outbred species seems unattainable.

Species inbreeding to some extent in nature

With organisms which practice a fair amount of inbreeding in nature, however, the problem is much less acute. Homozygous recessives will appear much more frequently than in random mating populations; if these homozygotes are harmful, they will tend to be eliminated by natural selection, so that harmful recessive alleles are removed from the population to a much greater extent than in outbreeding populations. Hence heterozygotes for harmful recessives are relatively infrequent. While this distinction between outbreeding and partially inbreeding populations is most marked in cases where the fitness of the recessive homozygote is low and where selfing is very common, the difference is quite noticeable in other cases. Consider as illustration the following very special case. At some locus, in a *very large* population, there are two alleles $\underline{A}$ and $\underline{a}$. Genotypes $\underline{AA}$ and $\underline{Aa}$ are equally viable, while $\underline{aa}$ has viability, relative to the viability of the other two genotypes $(1-k):1$ (in this, as in many other problems, only relative viabilities matter; we discuss this in detail in chapter 4). Write p, q for the frequencies of alleles $\underline{A}, \underline{a}$ respectively and define f as equal to

$$1 - \frac{\text{frequency of } \underline{Aa}}{2pq}$$

so that

$$\text{frequency of } \underline{Aa} = 2pq(1-f)$$

this frequency being measured among newly formed zygotes. Let $\underline{A}$ mutate to $\underline{a}$ at rate μ and ignore back mutation (this will not affect the result materially). Then it may be shown (Haldane 1940) that eventually q reaches a value

$$\frac{-f + \sqrt{f^2 + 4(1-f)\mu/k}}{2(1-f)}$$

and stays at that value thereafter. In a random mating population $f = 0$ and the formula just given reduces to $\sqrt{\mu/k}$, a formula which may well be familiar to the reader. In chapter 1, we showed that if s is the proportion of seed arising from selfing and $(1-s)$ the proportion arising from random mating, then in the absence of mutation or selection, the frequency of $\underline{Aa}$ finally settled down to

$$2pq\frac{2(1-s)}{2-s}$$

so that

$$1 - f = \frac{2(1-s)}{2-s}, \quad \text{whence} \quad f = \frac{s}{2-s}$$

and it may be shown that this formula still holds (near enough) in the present context. Suppose then as example that $\mu = 10^{-5}$ and consider two cases of mild selection, $k = 0.01$ and $k = 0.02$. Substitution in the above formulae yields the results given in table 2.1, where the percentage frequency of heterozygotes is given in the body of the table. The reduction in frequency resulting from selfing is really striking; with k as small as 0.01, the reduction is about 15-fold, as compared with random mating, for 50% selfing and about 47-fold for 75% selfing. For $k = 0.02$, the corresponding figures are 22-fold and 62-fold.

Table 2.1 Frequency of heterozygotes (per cent) among newly formed zygotes under some different mating systems for two regimes of mild selection against homozygous recessives (frequency of heterozygotes given in body of table).

	Mating system		
k	Random	50% selfing	75% selfing
0.01	6.12	0.40	0.13
0.02	4.37	0.20	0.07

The consequences are readily apparent. When species reproducing partly by selfing, partly by outcrossing are selfed, inbreeding depression is relatively slight. So striking is the contrast between outbreeding and partially inbreeding species in this respect that if we have a species of unknown mating system which shows little depression on inbreeding, we can conclude with confidence that our species practices a fair amount of inbreeding in nature. For example, two species of poppy, *Papaver rhoeas* and *Papaver dubium*, are often found intermingled in the same habitat. *P. rhoeas* is self-incompatible (but may be sibmated) whereas *P. dubium* can readily be selfed. Does it self much in nature? Ooi (personal communication) found that after one round of sibmating of *P. rhoeas* many plants appeared which were chlorophyll deficient, male sterile or female sterile. No such very harmful mutants appeared on inbreeding *P. dubium* from the same habitats, nor have they appeared after nine successive selfings of *P. dubium* from various populations. There is thus a strong presumption that this species practices much selfing in nature, as can be confirmed in other ways (Humphreys and Gale 1974; Gale, Rana and Lawrence 1974).

In view of the preceding, we shall confine ourselves for the rest of this chapter to discussing the effects of selfing in the absence of natural selection, with the understanding that while our results will often apply roughly to outbreeding species, they should be taken to apply with near-exactitude only for species which self to a fair amount in nature.

Effect of selfing on genotype frequencies at a single locus

We begin by considering one locus at a time. We have already noted Mendel's result which may be informally stated as "heterozygotes go down by a half every generation; a heterozygote yields equal numbers of either homozygote". Now it will be apparent that this informal statement is too brief to be exact. Suppose, say, that we started with eight individuals, two $\underline{AA}$, two $\underline{aa}$ and four $\underline{Aa}$. We self these individuals and raise just one offspring from every parent, thus keeping the number of individuals at eight in the next generation. If our statement held exactly, we should necessarily find the eight progeny individuals constituted as follows:

$$3\underline{AA}, \quad 3\underline{aa}, \quad 2\underline{Aa}$$

Of course this does not necessarily happen in practice. The numbers just given are "expected" numbers; observed numbers will differ from these owing to chance.

Experience indicates that to cope with this complication we need to be a little fussy with our definitions. However, we shall try to keep formality to a minimum. For example, we shall not use the term "probability", which is appropriate for discussion of problems involving chance, in this chapter. Rather we accustom the reader to chance processes by discussing an actual experiment which could be carried out quite easily in practice (perhaps with a little technical assistance) by the reader. It is hoped that this will serve to familiarize the reader with random processes and that when we introduce the term "probability" in the next chapter the concept will seem quite natural.

Imagine then an experimenter, who raises a large number of inbred lines of some species of flowering plant every generation (in principle, the number of lines should be "indefinitely large" but the reader will not go far wrong by imagining a constant large number of lines—say 2000—every generation). For simplicity, we suppose that in every generation every line is maintained by selfing and that just one offspring is raised from every parent plant (this restriction that every parent plant shall be represented in the next generation by just one offspring is not really necessary, but makes it easier to envisage the experiment). Notice that in any generation, every line is a Mendelian population size one (see chapter 1) and indeed it is possible to treat selfing as "random mating in a population of size one" (unnatural as this may sound). Since the number of lines (=populations) is indefinitely large, the proportion of lines heterozygous goes down by a half, *exactly*, every generation. Thus the statement that the "expected" proportion of heterozygotes is halved should be translated "if we have an indefinitely large number of lines, the proportion of lines heterozygous is halved". If the number of lines were small, the actual proportion obtained could, by chance, differ substantially from expected, but given 2000 lines the deviation will not be large in most cases. For the sake of precision, we shall throughout use the term "proportion of lines" to mean "proportion of lines when the number of lines is indefinitely large" but as already stated, no great harm is done by thinking of a definite large number of lines. Let H_0 be the proportion of lines heterozygous at a given locus at the start (that is, before we have done any selfing—generation 0) and H_n the corresponding proportion in generation n (that is, after n successive selfings). We have

$$H_1 = \tfrac{1}{2}H_0, \quad H_2 = \tfrac{1}{2}H_1 = (\tfrac{1}{2})^2 H_0$$

and generally

$$H_n = (\tfrac{1}{2})^n H_0$$

Clearly, if we think of any given locus, the large majority of lines will be homozygous after about 8 successive selfings, even if a large proportion of lines were heterozygous at the start. However, a line homozygous at some loci may still be heterozygous at others. Thus our formula does not supply the proportion of lines homozygous at all loci. It might be supposed that to calculate this latter proportion we should need to estimate the number of loci. Fortunately, such a hazardous procedure is not necessary; we discuss our requirements in the following brief digression.

Information needed to discuss the effect of inbreeding on the entire genotype

Our first requirement is the chromosome number; we write the haploid chromosome number as v. Our second is the *total* map length L, defined as $\tfrac{1}{2}$ the average number of chiasmata per nucleus. Map length so defined is measured in morgans (M); for example, if the average number of chiasmata is 34, the *total* map length is 17 morgans, or 1700 centimorgans (cM). Average chiasma counts for pollen mother cells may not be the same as for embryo sac mother cells; depending on the species concerned, the difference can be anything from zero to 100%, the latter occurring when meiosis is achiasmate in one type of sporogenesis. In principle, this is not a difficulty, since it will be apparent later that we just take the average of the two values. In practice, chiasma counts are more difficult to obtain for embryo sac mother cells than for pollen mother cells, so that a value may be known for the latter case only. Usually the difference between the two values does not exceed 25% and is often less (John and Lewis 1965) and, if so, we do not go too far wrong by using the pollen mother cell value, but in view of the possibility of achiasmate meiosis, as in *Fritillaria amabilis* (Noda 1968), some caution is needed. Similar sex differences are, of course, well known in animals.

What is the relation between map length and recombination fraction? Consider a pair of loci, $\underline{A}$ and $\underline{B}$ on the same chromosome. The map length between them is defined as $\frac{1}{2}$ of the average number of chiasmata between them. If $\underline{A}$ and $\underline{B}$ are sufficiently close together, two or more chiasmata between them will hardly ever be found; we either have no chiasmata or just one chiasma. If so, the map length between $\underline{A}$ and $\underline{B}$ must be

$[0 \times$ proportion of cells with no chiasmata between $\underline{A}, \underline{B}$

$+ 1 \times$ proportion of cells with 1 chiasma between $\underline{A}, \underline{B}]/2$

$= \frac{1}{2}$ the proportion of cells with 1 chiasma between $\underline{A}, \underline{B}$

Now by definition the recombination fraction between $\underline{A}, \underline{B}$ is the proportion of gametes recombinant between $\underline{A}, \underline{B}$. Moreover, in every cell in which one chiasma occurs between $\underline{A}, \underline{B}$, 2 of the 4 chromatids are recombinant, i.e. $\frac{1}{2}$ of the gametes eventually produced (or $\frac{1}{2}$ on average if we are thinking of female gametes) are recombinant. Hence for loci close together

recombination fraction $= \frac{1}{2}$ the proportion of cells with 1 chiasma
(as a decimal) $=$ map distance (in morgans)

For loci farther apart this equality no longer holds. Several formulae relating map distance to recombination fraction have been proposed, but none is completely satisfactory and, indeed, a universally applicable formula may well not exist. This difficulty, however, will not be a problem for us, since all we require is the *total* map length. This can be determined, as explained earlier, as $\frac{1}{2}$ the mean number of chiasmata per nucleus; no account need be taken of where the chiasmata occur—it will become apparent that the results we shall obtain are formulated in such a way as to be quite unaffected by interference or localization. The critical point for us is that a map distance of *one* morgan represents two chiasmata on average, that is, *one* "exchange point" (point where recombination has occurred) *on average* per gamete.

Since over short distances map distance is the same as recombination fraction, then if we have many loci close together A, B, C, D... which span the length of a chromosome, the map length for that chromosome is

recombination fraction between A, B + recombination fraction between B, C + recombination fraction between C, D + ...

and we obtain the *total* map length L by summing over all bivalents. For organisms where both methods are applicable, they give much the same answer (the genetical method will tend to give an underestimate, since known loci may not quite span all chromosomes, but in well-studied organisms this bias will be slight). Thus in the mouse (C. E. Ford, personal communication) the cytologically determined autosomal map length, when averaged over sexes, is about 13.25 M, whereas the genetically determined length is about 12.50 M; the latter is presumably a slight underestimate both for the reason given earlier and perhaps because data for the male, which has rather lower recombination fractions in general than the female, may be over-represented.

Equipped with a knowledge of chromosome number and total map length, we are ready to start. A first shot at the problem was made, as in so many other fields of genetics, by Haldane (in 1937). However, he was not able to make allowance for double crossing over. Later developments are mainly due to Fisher (1949, 1954, 1959, 1965) who developed his "theory of junctions" to deal with our problem. We deal here with selfing; the analysis here is that of Bennett (1953). We shall conclude by outlining some recent developments due to Franklin (1977).

The start (generation 0)

In general, we do not know to what extent our starting material is heterozygous. It is convenient to suppose that at the beginning every locus is heterozygous. Of course, this means that we shall obtain an exaggerated figure for the number of generations required for nearly all lines to become homozygous at every locus. However, as we shall see, the proportion of the genotype heterozygous is rapidly reduced in the early generations while it takes, relatively speaking, a fair time for the small proportion remaining heterozygous at a later stage to be reduced to zero in most lines. Hence the exaggeration will not be at all marked, except in cases where the starting material is homozygous at a large majority of loci. We represent a homologous pair of chromosomes at this initial stage in figure 2.1.

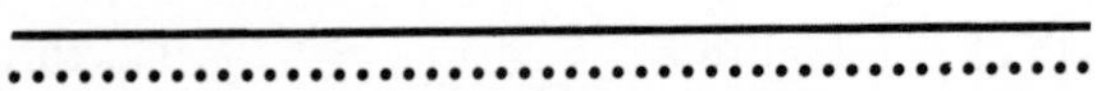

Figure 2.1. A homologous pair of chromosomes in generation 0. All loci heterozygous.

A later stage (generation n)

We now transfer our attention to some later generation, generation n say. If we concentrate on a single locus, we know that the proportion of lines heterozygous at this locus is

$$(\tfrac{1}{2})^n$$

and this is true for any locus we like. It follows that the proportion of the map length heterozygous, averaged over lines, is also

$$(\tfrac{1}{2})^n$$

and the average length heterozygous is

$$L(\tfrac{1}{2})^n$$

By this stage, parts of the chromosome pair will be heterozygous and parts homozygous. A particular case is illustrated in figure 2.2. In this particular case, there are three heterozygous "tracts" A, B and C, the remainder of the chromosome pair being homozygous.

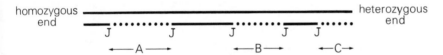

Figure 2.2. A homologous pair of chromosomes at a later generation. A, B and C are heterozygous tracts, the remainder is homozygous. J = external junction.

Proportion of lines completely homozygous at a late stage

Now consider not just one chromosome pair but all pairs in a given line. We count the number of tracts for every pair and add over all pairs to get the total number of tracts in a given line. When n is large, this total number of tracts will be few; tracts will be short and widely separated, and will therefore propagate independently. Under these conditions the proportion of lines with total number of tracts equal to some definite number, t say, is given by the *Poisson distribution* (see appendix 2). Suppose the total number of tracts, when averaged over lines, is m. Then from the Poisson distribution, the proportion of lines which have no tracts at all ($t = 0$) and are therefore homozygous at every locus is just

$$e^{-m}$$

(e being the usual mathematical constant $2.71828\ldots$). If we can find m, we can just look up e^{-m} in tables.

To determine the average number of tracts in a given generation

Examine figure 2.2. For tracts in mid-chromosome, we note that at each end of such a tract there is a point J at which heterozygous and homozygous regions meet. J is called an *external junction*.

Tracts at the end of the chromosome pair have J at one end and a heterozygous chromosome end at the other. Thus

$$\text{number of tracts} = \tfrac{1}{2}(\text{number of external junctions}$$
$$+ \text{number of heterozygous ends})$$

This result (which the reader can easily verify for figure 2.2) is, of course, true for any chromosome pair at any stage whatsoever. Adding tracts, external junctions and heterozygous ends for all chromosome pairs in a given line, we have for the line

$$\text{total number of tracts} = \tfrac{1}{2}(\text{total number of external junctions}$$
$$+ \text{total number of heterozygous ends})$$

Averaging over all lines, we have

$$m = \text{average total number of tracts}$$
$$= \tfrac{1}{2}(\text{average total number of external junctions}$$
$$+ \text{average total number of heterozygous ends})$$

Thus we have two things to find, "ends" and "junctions". The first is easy, the second a little more difficult.

To determine the average number of heterozygous ends

A heterozygous end must terminate in a heterozygous locus. Suppose $v =$ the haploid chromosome number, so that there are v pairs of chromosomes with 2 ends ("left" and "right") per pair. Thus we have $2v$ ends in all. Initially every end terminates in a heterozygous locus; the average proportion of such terminal loci still heterozygous in generation n is

$$(\tfrac{1}{2})^n$$

Hence the average number of ends heterozygous in generation n is

$$2v(\tfrac{1}{2})^n$$

To determine the average number of external junctions

Consider a tract, one morgan long in some generation, r say. In gametogenesis, one new exchange point, on average, will be formed somewhere in this region. A sample of such recombinant chromatids is shown in figure 2.3. When gametes unite to form a zygote, the chromosome pair will exhibit 2 new external junctions, on average, for the region (we ignore the very remote possibility that uniting gametes would have a new exchange point at exactly the same place). This also is shown in figure 2.3.

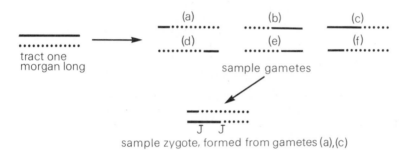

sample zygote, formed from gametes (a),(c)

Figure 2.3. Formation of new junctions. A tract one morgan long gives rise, on average, to 2 new junctions J one generation later.

For convenience, we have drawn our tract exactly one morgan long, but exactly the same argument would apply if we had a number of tracts whose length totalled one morgan. We conclude:

for every morgan of heterozygous tract present in generation r before gametogenesis, we get on average 2 new external junctions in generation $(r+1)$.

Now we showed earlier that the total length of tract, averaged over lines, in generation r is

$$L(\tfrac{1}{2})^r$$

Hence the number of new external junctions, averaged over lines, which first appear in generation $(r+1)$ zygotes is

$$2L(\tfrac{1}{2})^r$$

What happens to them later? Consider the situation in the *immediate neighbourhood* of a newly formed junction. One chromosome, *j* say, carries an exchange point, the other chromosome *e* does not. For a while, the junction continues to mark the boundary between two regions, one homozygous and one heterozygous. Eventually, however, the line will, as a result of further selfing, either become homozygous for the exchange point ("junction fixed") or homozygous for the corresponding short region in the chromosome which does not contain the exchange point ("junction lost"), as illustrated in figure 2.4.

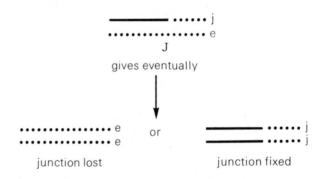

Figure 2.4. Ultimate fate of an external junction.

Thus the junction J continues to mark the boundary between a homozygous and heterozygous region until such time as it is lost or fixed. We see that the junction behaves like a mutation *j* first appearing in heterozygous condition

$$je$$

in generation $(r+1)$; "lost" *ee* and "fixed" *jj* are precisely analogous to the locus having become homozygous. Thus the average proportion of junctions first appearing in $(r+1)$ which are still extant as boundary markers in generation *n* (that is, $n-r-1$ generations later) is

$$(\tfrac{1}{2})^{n-r-1}$$

Thus the average number of external junctions which originated in gametogenesis in *r* and are still surviving as boundary markers in *n* must be

average initial number × average proportion surviving

$$= 2L(\tfrac{1}{2})^r \times (\tfrac{1}{2})^{n-r-1} = 2L(\tfrac{1}{2})^{n-1}$$

irrespective of the value of r. If we want surviving junctions in n, we must sum the expression just obtained over all generations prior to n from which these junctions might have arisen in gametogenesis; that is, all generations from 0 to $n-1$ inclusive, n generations in all. Hence

number of external junctions, averaged over lines, still acting as boundary markers in n is $2Ln(\frac{1}{2})^{n-1}$.

Proportion of lines homozygous at all loci

Adding junctions and ends we have finally

$m =$ number of heterozygous tracts averaged over lines

$\quad = \frac{1}{2}(\text{average external junctions} + \text{average heterozygous ends})$

$\quad = \frac{1}{2}[2Ln(\frac{1}{2})^{n-1} + 2v(\frac{1}{2})^{n}]$

$\quad = L(\frac{1}{2})^{n}\left[2n + \dfrac{v}{L}\right]$

As already stated, the proportion of lines homozygous at all loci is

$$e^{-m}$$

provided n is large enough for m to be small. For smaller n, the proportion of lines homozygous at all loci will be negligibly small.

An illustrative example

Results will depend to some extent on the species under discussion. In table 2.2 we give numerical results for *Papaver dubium*, for which L is about 17 (Humphreys, personal communication) and $v = 21$. The calculations are very simple; the reader can easily carry out similar calculations on his own favourite plant species.

Conclusions are so obvious as hardly to require any comment. Perhaps the main thing to notice is that, while it takes only 16 generations or so for homozygosity at all loci to be a near-certainty, the figure of 16 is rather larger than might be expected on the basis of what happens at a single locus. Thus in generation 8 only 32% of lines are homozygous at all loci, in spite of the fact that only $(\frac{1}{2})^{8} = 0.39\%$ of lines are still heterozygous at any given locus. From calculations rather too elaborate to be given here, the implication is as follows. By the time nearly all lines are homozygous at a given locus, say generation 8 in our case, the total length heterozygous is quite small in virtually all lines. For

Table 2.2 Mean number of tracts and proportion of lines homozygous at all loci following n successive selfings, for a species with 21 pairs of chromosomes and total map length 17 morgans.

Generation (n)	Mean number of tracts (m)	Proportion of lines homozygous at all loci
0	21.0000	
1	27.5000	
2	22.2500	
3	15.3750	
4	9.8125	
5	5.9688	
6	3.5156	0.03
7	2.0234	0.13
8	1.1445	0.32
9	0.6387	0.53
10	0.3525	0.70
11	0.1929	0.82
12	0.1047	0.90
13	0.0565	0.95
14	0.0303	0.97
15	0.0162	0.98
16	0.0086	0.99

$L = 17$, $v = 21$, $n = 8$, for example, the total length heterozygous in the whole genotype will very rarely exceed half a morgan. Much time, however, is taken up in removing the remaining heterozygosity in all lines, in that a smallish proportion of lines retain a little heterozygosity for quite a time. For example, we see from table 2.2 that the proportion of lines in which the last remaining tract does not disappear until we reach generation 13 is

$$0.95 - 0.90 = 0.05 = 5\%$$

Total length heterozygous

In the preceding section we indicated that, at a stage when most lines still retain some heterozygosity, it is possible to determine the proportion of lines whose total length heterozygous exceeds some stated value, say X morgans (Fisher 1965). This is helpful in enabling us to assess the effects of inbreeding at an intermediate stage of the process. However, we can obtain a rough idea of what is happening by merely considering the line-to-line variation in total length heterozygous. At an intermediate or late stage, the variance over lines of total length heterozygous is simply

$$2ma^2$$

where m, as before, is the average number of tracts and a is the average length of a tract in morgans (Fisher 1954; Bennett 1954). For the case of selfing

$$a = \frac{1}{2n + v/L}\text{morgans}$$

at this stage. Our simple formula for the variance is not, however, correct for the first few generations.

Franklin (1977) has shown that under selfing the variance in any generation n is

$$2(\tfrac{1}{4})^n \sum_{i=1}^{n} \frac{n!}{i!(n-i)!} \frac{1}{(4i)^2}\left(4iL - v + \sum_{j=1}^{v} e^{-4il_j}\right)$$

where l_j is the length in morgans of the jth bivalent. This formula was obtained subject to a number of assumptions which are discussed in detail by Franklin, who states that these assumptions are not critical to the outcome. The formula is a little awkward to use in practice, since it requires a knowledge, not just of total map length, but of map length of every bivalent. For an organism where it is possible to obtain map lengths from genetical data this is not a problem, but if map lengths of individual bivalents are to be obtained cytologically, individual bivalents must be recognizable. Such recognition is difficult in practice, being dependent on the extent to which individual chromosomes differ morphologically; the characteristic bands appearing after appropriate staining, which are so helpful for achieving recognition in mitosis, are not so clear in meiosis and are difficult to recognize where chromosomes are involved in chiasmata. However, rough estimates can be made. Of course, for n not too small, the variance should be near enough the same as given by Fisher and Bennett's formula. Franklin calculates the variance for inbreeding in maize by both formulae (but with a slight variant on Fisher and Bennett's value for a, which he puts equal to $1/(2n)$ and finds close agreement for generation six onwards.

Summary

Under inbreeding, the proportion of individuals heterozygous is reduced and the proportion homozygous increased. In outbreeding populations, harmful recessive alleles are present mainly in heterozygous form; when the corresponding homozygotes are formed under inbreeding, a marked reduction in fitness, known as inbreeding depression, is observed. In more inbred populations, however, such homozygotes will often have appeared in nature, so that harmful recessive alleles are kept at a very low frequency by natural selection. In these cases, little inbreeding depression is found.

The closer the degree of relationship of individuals mated, the more rapidly do the effects of inbreeding appear. Thus a given reduction in heterozygosity (accompanied by inbreeding depression in outbreeding species) will occur more rapidly under selfing than under any other system of inbreeding.

In cases where inbreeding depression (and hence natural selection during the inbreeding process) is marked, homozygosity is still reached eventually in many cases in practice, but theoretical analysis of the process is difficult. For cases where natural selection is not a complication, however, it is possible to calculate the time required for individuals to become homozygous at all loci. In the case of selfing, this analysis, although based on ideas of considerable subtlety and ingenuity, involves only simple algebra, and the final numerical calculations required to apply the analysis are very easy. It turns out that while the proportion of the genotype heterozygous is reduced very rapidly, being quite small after about 8 generations, it will take about 8 further generations of selfing to eliminate the residual heterozygosity in (virtually) all individuals.

CHAPTER THREE

RANDOM GENETIC DRIFT

Time and chance happeneth to them all

Ecclesiastes ix.11

On the word "probability"

Random changes are difficult to discuss, in that we cannot (usually) say that some specific change will happen, only that the change is probable or improbable. Statements involving probability are apt to produce discomfort in the reader, a vague sense of entering through a shop door labelled "only abstractions sold here". This discomfort will be dispelled if the reader makes a practice of rewriting statements involving the word "probability" into rather more cumbersome statements about *proportions*. Suppose, for example, we read, "If a woman aged 46 reproduces, the probability that her child will suffer from Down's syndrome is 1/25". This means: if an indefinitely large number of women aged 46 reproduce, 1/25 of all the children born to such women will have Down's syndrome.

Similarly, suppose we say "if we start with a population in which allele frequencies are $\underline{A}\frac{1}{2}, \underline{a}\frac{1}{2}$ and subsequently allele frequencies change as a result of drift only, then ultimately the population will end up either all $\underline{AA}$ or all $\underline{aa}$; the probability that it ends up all $\underline{aa}$ is $\frac{1}{2}$". This means: take an indefinitely large number of populations all with allele frequencies $\underline{A}\frac{1}{2}, \underline{a}\frac{1}{2}$ and let all the populations run for a very long period of time, allele frequencies changing as a result of drift only; in the end $\frac{1}{2}$ of the populations end up all $\underline{AA}$, $\frac{1}{2}$ of the populations end up all $\underline{aa}$. Whenever we discuss probability, we shall have some such notions in mind.

Intuitive notions on the effects of drift

We now propose some rather intuitive ideas about drift. Bearing in mind Feller's dictum "natural intuition and natural thinking are a poor affair", we shall sketch a mathematical verification of most of these ideas later.

Consider a population, size N, with just one $\underline{A}$ allele, the other $(2N-1)$ alleles at the locus being $\underline{a}$. Then the individual carrying the $\underline{A}$ must be heterozygous $\underline{Aa}$. For reasons quite unconnected with his genotype at the $\underline{A}/\underline{a}$ locus, he may not reproduce. Even if he does reproduce, he may not transmit the $\underline{A}$ to any of his offspring, just by chance. If so, the $\underline{A}$ has been lost from the population, purely by chance, *even if the possession of $\underline{A}$ conveys a selective advantage.* If $\underline{A}$ is lost, it can reappear only as a result of new mutation—this may not happen for a long time. Moreover, if $\underline{A}$ is not lost in one generation, it may be lost later.

The same argument applies to a lesser extent, even if there is more than one $\underline{A}$ in the population, as long as $\underline{A}$ is very rare. Similarly, if $\underline{A}$ is very common, we can apply the argument to the very rare $\underline{a}$.

We conclude that if p, the frequency of an allele such as $\underline{A}$, is near 0 or 1, drift can have an important effect on what happens. This is true even if the population is large and whether or not the allele conveys a selective advantage.

When p is not near 0 or 1, "compensation" will occur, that is some $\underline{Aa}$ will have fewer offspring carrying $\underline{A}$ than expected, but some will have more. Some $\underline{AA}$ will have fewer offspring than average, some will have more. Opportunities for this compensation will be more marked in large populations than small. While it would be wrong to suppose that compensation is ever complete, we can conclude, tentatively, that with p not near 0 or 1, drift will bring about rather smallish changes, at least in large populations, provided we consider only short periods of time.

Nevertheless, we have to face the possibility that the effect of these small changes is cumulative over time; after a very long time, the population might end up all $\underline{AA}$ or all $\underline{aa}$. If, at the start, $\underline{A}$ is fairly rare, we are more likely to end up all $\underline{aa}$, *other things being equal.* In order to study the process of drift, we shall in the first instance suppose such equality. In particular, we shall assume that $\underline{AA}$, $\underline{Aa}$ and $\underline{aa}$ are all equally fit—we say that $\underline{A}, \underline{a}$ are *neutral genes* (more correctly, neutral alleles). We shall bring in natural selection *later* (in chapter 4). We also, for simplicity, assume random mating.

Probability distribution of allele frequency

Suppose then that we have a population size N. For simplicity, assume just two alleles $(\underline{A}, \underline{a})$ at a locus. We write p_0 for the initial (generation 0) frequency of $\underline{A}$ and q_0 for the initial frequency of $\underline{a}$, with $p_0 + q_0 = 1$.

To see the effects of drift we can ask: what is the probability distribution of p, the frequency of allele $\underline{A}$, t generations later? That is, consider an indefinitely large number of populations, all the same size N and all with frequency of $\underline{A}$ equal to p_0 initially. In generation t we examine these populations and for each find the frequency of allele $\underline{A}$. We ask: in what proportion of these populations is the frequency 0, in what proportion is the frequency $1/(2N)$, in what proportion is the frequency $2/(2N)$, and so on.

This probability distribution was determined by Kimura in 1955 (his result is slightly approximate, but good enough unless N is very small). However, the derivation is difficult. Fortunately, we can get the essence of the matter by considering some properties of this distribution, which can be obtained, without too much difficulty, without knowing the distribution itself; we shall find the latter for special cases only (for example, when t is very large).

Mean allele frequency: mean frequency of heterozygotes

We might surmise that, if we knew the mean and variance of our unknown distribution, we should make some progress and this is in fact so. We shall determine the mean, but instead of the variance we shall find in the first instance a closely related but more easily understood quantity in the present context, namely the mean frequency of heterozygotes. Writing p for the frequency of allele $\underline{A}$ in some given generation we write Ep for the mean ($=$ expected) value of p. The mean ($=$ expected) frequency of heterozygotes is

$$E[2p(1-p)] = 2E[p(1-p)] = 2Ep - 2Ep^2$$

As a matter of definition (see appendix 2), the variance of p is

$$E[p - Ep]^2 = Ep^2 - [Ep]^2$$

so that

$2 \times$ the variance $= 2 \times$ the mean $- 2$ (the mean)^2

$-$ mean frequency of heterozygotes

Suppose we have a population of N individuals ($2N$ alleles) with allele frequencies $\underline{A}\,p, \underline{a}\,q$ ($p + q = 1$). The next generation will be a sample of $2N$ alleles drawn at random from an "infinite" ($=$ indefinitely large in size) population of possible alleles with frequencies $\underline{A}\,p, \underline{a}\,q$.

Start off in generation 0 with an indefinitely large number of populations, each size N and each with allele frequencies $\underline{A}\, p_0$, $\underline{a}\, q_0$. In the next generation, the frequency of $\underline{A}$ will vary with population; for any population, this next generation will be a sample of $2N$ alleles drawn at random from an infinite population of possible alleles with frequencies p_0, q_0. We have one such sample from every parent population, all parent populations being the same. Thus we can regard the indefinitely large number of population in generation 1 as if they were drawn as random samples, size $2N$, from one infinite population of possible alleles with frequencies p_0, q_0. Hence the probability of getting r $\underline{A}$ alleles in generation 1 ($=$ the proportion of populations with exactly r $\underline{A}$ alleles in generation 1) is given by the *binomial* distribution (readers unfamiliar with this distribution may care to consult appendix 2); this being the probability distribution of allele *numbers* in generation 1. This binomial distribution has

$$\text{mean } Er = 2Np_0, \quad \text{variance } V(r) = 2Np_0q_0$$

from the standard properties of the binomial distribution. Allele frequencies are, of course, obtained by dividing allele numbers by $2N$; hence to find the mean and variance of the distribution of allele *frequencies*, we must divide Er and $V(r)$ by $(2N)$ and $(2N)^2$ respectively. Hence the probability distribution of allele frequencies in generation 1 has

$$\text{mean} = p_0, \quad \text{variance} = \frac{p_0q_0}{2N}$$

and mean frequency of heterozygotes is

$$2(\text{mean}) - 2(\text{mean})^2 - 2(\text{variance})$$

$$= 2p_0 - 2p_0^2 - 2\frac{p_0q_0}{2N}$$

$$= 2p_0q_0\left(1 - \frac{1}{2N}\right)$$

since $p_0 + q_0 = 1$.

Thus, while the mean frequency of $\underline{A}$ in generation 1 is p_0, the mean frequency of heterozygotes has gone down by a factor

$$\left(1 - \frac{1}{2N}\right)$$

This rather suggests that the mean frequency of $\underline{A}$ in any generation will be p_0, whereas the mean frequency of heterozygotes declines by a factor $(1-1/2N)$ per generation, as can indeed be proved rigorously (see appendix 2). Hence, if we look at the complete set of populations in some later generation t (say), we shall find that the frequency of allele $\underline{A}$ will vary from one population to another; if we average these allele frequencies we shall get

$$\text{mean allele frequency} = p_0$$

Similarly, the frequency of heterozygotes will vary; averaging these frequencies, we shall find

$$\text{mean frequency of heterozygotes} = 2p_0q_0\left(1 - \frac{1}{2N}\right)^t$$

a result due to Wright (1931).

Extreme versus intermediate allele frequencies

We are now in a position to examine more rigorously the points that we set down intuitively earlier. Consider first the relative importance of drift when p is near 0 or 1, compared with the case when p is intermediate. To do this, consider a population with size N, allele frequency of $\underline{A}$ equal to p in some generation t, and consider all populations to which it *might* give rise one generation later (generation $t+1$). Our actual population one generation later can be regarded as drawn at random from such a set of potential populations. Out of all these potential populations, the proportion of populations in which there will be just r $\underline{A}$ alleles is given by the binomial distribution, with

$$\text{mean } 2Np, \quad \text{variance } 2Npq$$

where $q = 1-p$. Suppose N is not very small. Then for intermediate p we may (following a standard result) approximate this binomial distribution by a normal distribution with

$$\text{mean } 2Np, \quad \text{variance } 2Npq$$

Hence we may approximate the distribution of frequency of allele $\underline{A}$ by a normal distribution

$$\text{mean } p, \quad \text{variance } \frac{pq}{2N}$$

Thus, when N is large, the variance is very small, so that we have a normal distribution with a very narrow spread around the mean p; hence in almost all potential populations the allele frequency has hardly changed at all from its previous value p. Recalling that our actual population in $(t+1)$ can be regarded as drawn at random from the set of potential populations, we see that in practice changes in allele frequency in one generation due to drift in a large population will usually be very small.

When p is small, the normal approximation is poor. However, in such a case we may approximate the binomial by the Poisson distribution, so that the proportion of our potential populations with no $\underline{A}$ alleles at all will be about

$$e^{-2Np}$$

(e being the usual mathematical constant 2.71828...) and this can be quite large if p is small. For example with just one $\underline{A}$ allele in generation t, so that $p = 1/2N$ we have

$$e^{-2Np} = e^{-1}$$

which (from tables) equals 0.37.

We have shown then that for largish populations, drift has virtually no effect on intermediate allele frequencies in a single generation, but that very rare alleles may be extinguished by drift very readily.

On the other hand, it is a universal experience that very small samples often differ quite markedly from the parent population from which they are drawn. This being so, drift will have a marked effect on the genetic composition of very small populations; quite striking changes in allele frequency could occur in just one generation.

Circumstances under which drift may be ignored

The reader may well protest at this stage at our use of the rather vague terms "largish" and "very small". Obviously, we need some more precise criterion which will enable us to identify situations in which the effect of drift is unimportant. Now even when the population size is very large, drift in one generation has some effect (although very small). If effects cumulate over time, then very small effects in a single generation can lead to large effects over a very long period of time. Such cumulative changes are most readily studied by following the change in mean frequency of heterozygotes. In generation t, this frequency is

$$2p_0q_0\left(1 - \frac{1}{2N}\right)^t$$

or about

$$2p_0q_0e^{-t/2N} \quad \text{(see appendix 2)}$$

How many generations are required for this to fall to, say, $\frac{1}{2}$ of the initial heterozygote frequency $2p_0q_0$? We put

$$2p_0q_0e^{-t/2N} = \tfrac{1}{2}2p_0q_0$$

so that

$$e^{-t/2N} = \tfrac{1}{2}$$

whence

$$-t/2N = \log_e \tfrac{1}{2} = -0.693 \quad \text{(from tables)}$$

giving

$$t = 1.386N \text{ generations}$$

a very long time unless N is small. The reader is recommended to repeat this calculation for values other than $\frac{1}{2}$.

Our discussion illustrates a very helpful rule when thinking about effects of drift, namely "apart from very unusual chance exceptions, changes due to drift will be very small if the number of generations involved is very much less than the population size". As a practical guide, we can take "very much less" to be less than one hundredth. A little care is needed when using this rule since, as we have repeatedly stressed, small changes in frequency may be critical for rare alleles. With this qualification, we can apply the rule very freely. For example, if we have a *Drosophila* cage population, of size 3000 or so, it follows at once that only very small changes in allele frequency will occur in, say, 10 generations.

To verify this rule, we recall that

$$2 \times \text{variance} = 2 \times \text{mean} - 2 \text{ mean}^2 - \text{mean frequency of heterozygotes}$$

so that if in generation t we find the frequency of allele $\underline{A}$ in each of our indefinitely large number of populations, the variance of these allele frequencies is approximately

$$p_0 - p_0^2 - p_0 q_0 e^{-t/2N} = p_0 q_0 (1 - e^{-t/2N})$$

A very small value of this variance implies that in almost all populations, the frequency of $\underline{A}$ is very close to the mean frequency p_0. Since the latter is also the initial frequency, a very small variance means that, in nearly all populations, allele frequencies have hardly changed from their initial value. If t is very much less than N, $e^{-t/2N}$ will be very close to unity, so that the variance will be close to zero. For example, if t is less than $N/(100)$ we find from tables that

$$e^{-t/2N} \text{ is greater than } 0.995$$

so that the variance is less than $p_0 q_0/200$.

Long-term effects of drift

Notice that as t gets larger and larger

$$\left(1 - \frac{1}{2N}\right)^t \to 0$$

(the reader may care to try this for a few values of N and t), so that ultimately the mean frequency of heterozygotes becomes zero. Since the mean of a set of non-negative quantities can be zero only if all the quantities are zero, this means that ultimately there are no heterozygotes at all in any of our populations; populations are of two kinds only, either all $\underline{AA}$ ($p = 1$) or all $\underline{aa}$ ($p = 0$)—we refer to the former as "$\underline{A}$ fixed" and to the latter as "$\underline{A}$ lost". Suppose the proportion of populations which are all $\underline{AA}$ is α and the proportion of populations which are all $\underline{aa}$ is β. We have then the ultimate probability distribution of p, the allele frequency of $\underline{A}$, made up of $p = 0$ with probability β, $p = 1$ with probability α; all other values of p have probability zero. By definition of a mean, the mean of this ultimate distribution is

$$0 \times \beta + 1 \times \alpha = \alpha$$

and this must equal p_0, the initial frequency of $\underline{A}$, since the mean is p_0 at any time, including ultimate time. We have therefore the simple result

probability of fixation = frequency at the start

For example, if we start with just one $\underline{A}$ allele, newly arisen by mutation in the population, all other alleles at the locus being $\underline{a}$, then the probability that the population will end up all <u>AA</u> is

$$p_0 = \frac{1}{2N}$$

(Fisher 1930, Wright 1931). In terms of our indefinitely large number of populations, a proportion $1/(2N)$ ends up all <u>AA</u> ($p = 1$), a proportion $(1 - 1/(2N))$ ends up all <u>aa</u> ($p = 0$). Thus the newly arisen neutral allele is lost in almost all populations; its chance of fixation is very small unless the population size is very small.

Variable fertilities

It will probably be apparent that if fertilities are very unequal, with rather few individuals giving rise to most of the next generation, the population will behave in regard to drift as if its size were considerably smaller than its actual size N. More generally, the size will effectively be less than the actual unless all individuals have an equal capacity to contribute to the next generation.

Wright has shown how this complication can be overcome. The basic idea is to calculate the "effective population size" N_e and use the formulae given earlier in this chapter, replacing N by N_e *whenever this is appropriate* (see below; the reader is warned against an uncritical substitution of N_e for N in every formula). For an extensive discussion, see Crow and Kimura (1970); we merely quote a few results.

Suppose that the adult population size N is constant. Let g be the number of gametes contributed by a parent to the next adult generation; g will vary from parent to parent. Since the population is of constant size, the mean of the gs will be 2; write $V(g)$ for the variance of the gs. Then it turns out that

$$N_e = \frac{4N - 2}{2 + V(g)}$$

Suppose all individuals are equally likely to contribute a given number of gametes to the next adult generation (the actual number contributed will, owing to chance, vary from one individual to another). Then it may be shown that

$$V(g) = 2\left(1 - \frac{1}{N}\right)$$

so that $N_e = N$ and the theory given earlier in this chapter stands unmodified.

However, in practice, large differences between individuals in reproductive capacity are often found. This is certainly true in the case of flowering plants, where, as Stebbins (1971) has noted, seed number is subject to very great phenotypic modification depending upon the environment; thus very local differences in, say, soil fertility or crowding within the habitat will bring about a substantial variation in seed output (see also Levin (1978)). Crow and Morton (1955) have shown that N_e is decidedly less than N in *Drosophila*, *Lymnaea* and Man. We may surmise therefore that, in at least a substantial proportion of species, effective population size will be smaller than N.

It is, however, easy to allow for this. For example, the mean frequency of heterozygotes in generation t is obtained by taking our previous formula for the mean frequency of heterozygotes in generation t, and replacing N by N_e. The number of generations required to halve the mean frequency of heterozygotes is, therefore, $1.386N_e$. The probability of fixation, however, remains at p_0; thus the chance of ultimate fixation of a newly arisen neutral mutant allele is $1/(2N)$ and *not* $1/(2N_e)$. The time taken until fixation occurs, however, does depend on N_e. Kimura and Ohta (1969) have shown that, considering only those newly arisen neutral mutants which are ultimately fixed, the average time until fixation of such a mutant is approximately $4N_e$ generations. The proof of this result, although very pleasing, is rather too long to be given here.

Variable population size

So far we have taken the population size as constant. In practice, however, the population size will often fluctuate. Suppose the population is sometimes large, sometimes small. In regard to drift, does the population behave like a large population, like a small population or like a medium-sized population?

For simplicity, consider only cases where all individuals are equally likely to contribute a given number of gametes to the next adult generation (other cases give essentially similar results). Wright (1939) has shown that if we start in generation 0 with allele frequencies p_0, q_0 and if in successive generations $0, 1, 2, \ldots, t$ the population size takes values $N_0, N_1, N_2, \ldots, N_t$, the mean frequency of heterozygotes in generation t is approximately

$$2p_0q_0\left(1 - \frac{1}{2N^*}\right)^t$$

where N^* is the "harmonic mean" of the Ns (excluding N_0); that is,

$$\frac{1}{N^*} = \frac{1}{t}\left(\frac{1}{N_1} + \frac{1}{N_2} + \ldots + \frac{1}{N_t}\right)$$

Thus the population behaves as though it had constant size N^* (of course, if the population has constant size N, then $N^* = N$). If, with Wright, we substitute values of $N_1, N_2, \ldots, N_t$ in this formula for N^*, we find that if the population is sometimes large, sometimes small, N^* comes out rather small; thus the population behaves like a small population in regard to speed of change. For example, if $N_1 = 10$, $N_2 = 10^2, \ldots, N_6 = 10^6$, then N^* turns out to be 54. The probability of fixation, however, is still p_0; thus the chance of ultimate fixation of a newly arisen mutant allele, first appearing in generation 0 is $1/(2N_0)$ and *not* $1/(2N^*)$. Finally, consider the mean fixation time for such a mutant. Suppose that the population size changes in a cyclical manner, a given sequence of population sizes, say $N_1, N_2, \ldots, N_t$ being endlessly repeated. Let N^* be the harmonic mean of the population sizes, $N_1, N_2, \ldots, N_t$ for a single cycle. We should rather expect, from our previous work, that the mean fixation time would be about $4N^*$ and, provided that the number of generations in a single cycle is fairly small, this surmise proves to be correct (Chia 1968).

Summary

Frequencies of neutral alleles will change as a result of random genetic drift but, unless the population is small, changes will be very minor over short periods of time (i.e. if the number of generations is very much less than the population size). Slight changes in frequency of rare alleles, however, can lead to their extinction. Given a sufficiently large number of generations, drift will eventually lead to the fixation of just one allele. The rarer an allele initially, the less is the chance that it will ultimately be fixed.

CHAPTER FOUR

SURVIVAL AND FIXATION OF
ADVANTAGEOUS MUTANTS

Slave, I have set my life upon a cast,
And I will stand the hazard of the die:
I think there be six Richmonds in the field;
Five have I slain today, instead of him:

William Shakespeare, *King Richard III*

Factors affecting the chance of fixation

We have shown that a newly arisen neutral mutation has very little
chance of survival and fixation, unless the population size is very small.
Suppose, however, that the mutant conveys a selective advantage to its
possessor. Is the chance of survival and fixation greatly enhanced? Can a
selective advantage turn a rank outsider into the favourite?

The reader will probably surmise that if the selective advantage is
really large, comparable say with the advantage enjoyed by a melanic
moth in a smoke-polluted area, the prospects for the mutant allele would
be very bright. For mutants conveying a small advantage, however, it is
not so easy to decide what will happen. In the early days of the
development of population genetics, there was some uncertainty among
biologists as to how much natural selection could accomplish; indeed,
Fisher wrote his celebrated work *The Genetical Theory of Natural
Selection* in 1930 with the main aim of demonstrating the remarkable
power of natural selection (Fisher, personal communication). Fisher
(1930) and Wright (1931) using different methods, obtained a formula,
approximate but very accurate when the selective advantage is small,
giving the probability that a newly arisen advantageous mutant allele
will ultimately be fixed. A special case of this formula had been found by
Haldane in 1927, using yet another method.

44

However, when attempting to determine the prospects for an advantageous allele, we should prefer not to confine ourselves to the case where the allele has just arisen. In particular, the allele might have been neutral or even disadvantageous in the past but have persisted in the population, usually at low frequency, as a result of mutation and drift. Suppose now that the environment changes in such a way that our allele becomes advantageous. What happens?

We stress a point tacitly assumed in the previous chapter, that when we make a prediction we must do so in the light of the information we have; if the information were different, the prediction might be different. An observer at Bosworth, asked to assess Richard III's prospects, would no doubt have given a very different answer after Stanley's desertion than before. This being so, we must decide on our starting point and calculate the probability of fixation from there; we begin with our advantageous allele having a definite initial frequency p_0, in the same way as we did for a neutral allele in the previous chapter. We can see that the value of p_0 chosen will be critical, for the following reason. An advantageous allele may, just like a neutral allele, be lost by drift, but now this tendency to be lost is opposed by natural selection. Which "wins", drift or selection? Clearly, the greater the selective advantage and the weaker the effect of drift, the greater the chance of fixation. Now the chance of loss owing to drift is much greater for a rare allele than for a common one. Suppose an advantageous allele initially very rare manages to survive these early stages and increases substantially in frequency. The effect of drift is now very feeble unless the population is small. We surmise then (but will make the point more precisely later) that once the allele frequency has become substantial, ultimate fixation is almost certain, save in cases where both population size and selective advantage are quite small.

Thus we must decide on an appropriate value for p_0. If our allele was not pre-existent in the population but has just arisen by mutation, it is most informative to take this as the starting point, in which case

$$p_0 = \frac{1}{2N}$$

If the allele had previously been present in the population but had been neutral or disadvantageous and then, owing to a change in environment, has become advantageous, we take our initial frequency to be the frequency prevailing at the time when the allele became advantageous; usually this frequency will be small. It should not give cause for concern that we get one answer for the probability of fixation if we take p_0 to equal a small value, such as $1/(2N)$; whereas a different investigator, who looks at the population many generations later and notices that the allele frequency has risen to, say, 0.5, takes this as his initial frequency and gets a different answer from us for the probability of fixation. The point is that in making our prediction we were by no means certain (given possible effects of drift on a rare allele) that the frequency would ever reach 0.5 and had to incorporate this uncertainty into our calculations, whereas the later investigator could take the rise to 0.5 as given, since, although such a rise will not always happen, it has happened in his particular case.

Clearly, we need a formula which will give the probability of fixation of an advantageous allele, starting with an initial frequency p_0; in most applications p_0 will be small. Such a formula was obtained by Kimura in 1957. From the preceding discussion it follows that this formula will involve the factors relevant to drift, namely p_0 and the population size N (or N_e in cases where N_e differs from N) and, of course, the selective advantage, which we must now define in a more exact manner than hitherto.

When comparing the fitness of different genotypes, we must take account of differences both in survival to maturity and in fertility (in the broad sense, thus including capacity to obtain a mate). Survival is easy to represent. A newly formed zygote of defined genetic constitution situated in a specific environment will have a definite probability of survival to maturity. This probability is called the *viability*. Let the viabilities of genotypes AA, Aa and aa be u, v, w respectively. Then (ignoring random effects for the moment) we have:

	AA		Aa		aa	Total
Zygotic frequency	P		$2Q$		R	1
Viability	u		v		w	
Relative frequency at maturity	uP	:	$2vQ$	:	wR	T
Absolute frequency at maturity	$\dfrac{uP}{T}$		$\dfrac{2vQ}{T}$		$\dfrac{wR}{T}$	1

where we have divided the relative frequencies by their total T $(=uP+2vQ+wR)$ in order to obtain absolute frequencies adding to unity. It will be apparent that these absolute frequencies are unaffected if we divide the three viabilities by the same constant; in other words, if we are interested in calculating genotype (or allele) frequencies, only relative viabilities matter. This being so, we may choose one genotype as standard and divide all viabilities by the viability of the standard genotype which now has standardized viability (often written "viability" for brevity) unity. In the present chapter, we let a̲a̲ be the original wild-type homozygote and take this as standard. We write the (standardized) viability of the mutant homozygote A̲A̲ as

$$1 + s$$

where s will be positive since we are considering the case where the mutant confers a selective advantage; s is called the *selective advantage* of A̲A̲ over a̲a̲. The (standardized) viability of the heterozygote is written

$$1 + sh$$

sh being the selective advantage of A̲a̲ over a̲a̲. If A̲a̲ is as viable as A̲A̲ ($h = 1$), we say that A̲ is dominant in viability to a̲. If A̲a̲ is only as viable as a̲a̲ ($h = 0$), A̲ is recessive in viability to a̲. Otherwise, we confine ourselves *here* to cases where A̲a̲ is intermediate in viability to A̲A̲ and a̲a̲, h lying between 0 and 1, with $h = \frac{1}{2}$ representing the case of no dominance in viability. Note that we have tacitly assumed that relative viabilities at the A̲/a̲ locus are unaffected by differences between individuals at other loci. An assumption of this kind can, in some circumstances, give very misleading results (Lewontin 1974). Unless, however, such interactions with other loci are very marked, our simple model should not lead us seriously astray in the present context.

Differences in fertility are more difficult to represent. Any specified genetically distinct type of mating will, in general, have a fertility peculiar to itself. This leads to complications, in that the average fertility of a given genotype will depend on the frequencies of other genotypes present; thus genotype frequencies depend on fertilities which in turn depend on genotype frequencies. To make the problem tractable, Bodmer (1965) suggested that we assume (as a first approximation) that the *relative* fertilities, say f, g, l of our three genotypes in a single sex remain the same irrespective of the genotype of the individual with whom they mate. Data are too scanty for a definite decision to be reached as to whether this assumption is, in general, reasonable, although not surprisingly it does not always hold (Prout 1971). Even with the assumption of constant relative fertilities, it seems likely that the difference in fertility between a pair of genotypes will vary markedly with the sex of the individual concerned; the reader interested in the effects of these sex differences is referred to Bodmer's paper.

Only if we assume that relative fertilities are the same in both sexes can we fit fertility into the model we described earlier. We simply replace u, v, w by uf, vg, w and standardize as before; the standardized fitnesses can then be represented as $1+s$, $1+sh$, 1. We shall not make this very shaky assumption, but confine ourselves at this stage to discussing mutants which do not affect fertility or in which the effect on fertility is very minor compared to the effect on viability.

Genic selection

We have suggested that, under a wide range of circumstances, the whole question of whether an advantageous allele will survive and be fixed is settled during the stages when the allele is rare. But when the allele is rare, individuals homozygous for the allele will be very rare; the allele will be represented almost entirely in heterozygotes. In that homozygotes are so rare, it seems implausible that the chance of survival and fixation of our allele $\underline{A}$ should depend to any great extent on the viability of $\underline{AA}$ homozygotes. We should rather expect that the critical factor is the viability $(1+sh)$ of the heterozygote and this surmise turns out, as we shall see later, to be correct under a wide range of circumstances.

This enables us to introduce a helpful simplification, with the advantage of unifying the treatment of diploids and haploids. Suppose we have a population of, say, *Chlamydomonas* and for simplicity imagine that individuals are raised in such a way that mating is synchronous. Let the frequencies of genotypes $\underline{A}, \underline{a}$ immediately after meiosis be p, q. Write the relative fitnesses (including viability, capacity for asexual reproduction) of these genotypes up to the next round of mating as $1+\alpha, 1$. We have (ignoring random effects for the moment)

	$\underline{A}$	$\underline{a}$	*Total*
Frequency just after meiosis	p	q	1
Fitness	$1+\alpha$	1	
Relative frequency just before mating	$(1+\alpha)p \quad :$	q	T
Absolute frequency just before mating	$\dfrac{(1+\alpha)p}{T}$	$\dfrac{q}{T}$	1

where $T = (1+\alpha)p+q = 1+\alpha p$

Now consider diploids but think of the viability of *alleles* rather than of genotypes. By analogy with haploids, we write

	A	a
Frequency among zygotes	p	q
Viability	$1+\alpha$	1

and (ignoring random effects) obtain the frequency of the A allele just before the next round of mating as

$$p^* = \frac{p(1+\alpha)}{1+\alpha p}$$

the same formula as for haploids. When using this approach, which was introduced by Fisher (1922) we are said to be discussing "genic selection" or using the "haploid model".

What does this mean in terms of viability of zygotes? Suppose that the substitution of any A allele for any a allele *multiplies* the viability by $(1+\alpha)$—"uniform genetic selection". Under random mating (ignoring drift) we have

	AA	Aa	aa
Zygotic frequency	p^2	$2pq$	q^2
Viability	$(1+\alpha)^2$	$(1+\alpha)$	1
Absolute frequency just before mating	$\dfrac{p^2(1+\alpha)^2}{S}$	$\dfrac{2pq(1+\alpha)}{S}$	$\dfrac{q^2}{S}$

where $S = p^2(1+\alpha)^2 + 2pq(1+\alpha) + q^2$

$\quad = [p(1+\alpha)+q]^2$

$\quad = (1+\alpha p)^2$

The frequency of allele A just before mating is

$$p^* = [p^2(1+\alpha)^2 + pq(1+\alpha)]/S$$
$$= p(1+\alpha)[p(1+\alpha)+q]/S$$
$$= \frac{p(1+\alpha)}{1+\alpha p}$$

Thus genic selection is equivalent to uniform genetic selection, in that both give the same value for p^*, provided the selective advantage of allele A in genic selection is equated to the selective advantage of Aa in uniform genetic selection. Hence we have $\alpha = sh$. Now the viability of AA, $1+s$, will not usually equal $(1+\alpha)^2$ but this hardly matters since the viability of $\underline{AA}$ is, in many circumstances, irrelevant. Thus the haploid model will be a good enough approximation in these cases. Note that when α is small, α^2 is very small and may be neglected, so that we have, near enough, the viabilities

$\underline{AA}$	$\underline{Aa}$	$\underline{aa}$
$1 + 2\alpha$	$1 + \alpha$	1

that is, no dominance. Thus the haploid model should be particularly appropriate when dominance is absent or very incomplete.

To sum up this section: under a wide range of circumstances, we can discuss survival and fixation solely in terms of viabilities of alleles, $(1 + \alpha)$ for $\underline{A}$ and 1 for $\underline{a}$, where α is the selective advantage of the heterozygote. If p_0 is the frequency of $\underline{A}$ among the newly formed zygotes of generation 0, then the frequency of $\underline{A}$ just before the adults, resulting from these zygotes, mate is

$$p_0^* = \frac{p_0(1 + \alpha)}{1 + \alpha p_0}$$

provided we ignore drift. We consider the effect of drift in the next section.

Mean and variance of the change in allele frequency

Initially (generation 0, just after fertilization) the frequency of $\underline{A}$ is p_0. For neutral alleles, we found the probability distribution of the proportion of $\underline{A}$ alleles one generation later by drawing random samples of size $2N$ from an infinite population with frequencies p_0, q_0. We follow a similar procedure here, but now our infinite population has frequencies $p_0^*, q_0^* \; (= 1 - p_0^*)$ giving for generation 1, just after fertilization

$$\text{mean frequency of } \underline{A} = p_0^*$$
$$\text{variance of frequency of } \underline{A} = \frac{p_0^* q_0^*}{2N}$$

Consider our set of potential populations in generation 1. As in the case of neutral alleles, our actual population in generation 1 can be regarded as drawn at random from the set of potential populations. For the ith potential population, write δp_i for the change in frequency, that is

$$\delta p_i = (\text{frequency in generation 1}) - p_0$$

Then the mean of the values of δp_i, written $E(\delta p_i)$ is

$$p_0^* - p_0 = \frac{p_0(1+\alpha)}{1+\alpha p_0} - p_0$$

Now (appendix 2)

$$1/(1+\alpha p_0) = (1+\alpha p_0)^{-1} = 1 - \alpha p_0 + \text{terms in } \alpha^2, \alpha^3, \alpha^4, \ldots$$

When α is small, terms in $\alpha^2, \alpha^3, \alpha^4, \ldots$, will be negligibly small and we have the good approximation

$$\begin{aligned}
E(\delta p_i) &= p_0(1+\alpha)(1-\alpha p_0) - p_0 \\
&= p_0 + \alpha p_0 - \alpha p_0^2 - \alpha^2 p_0^2 - p_0 \\
&= \alpha p_0 q_0
\end{aligned}$$

if we ignore the term in α^2 and remember that $q_0 = 1 - p_0$.

Now consider the variance of the values of δp_i. We obtain a δp_i by taking the frequency of $\underline{A}$ in the ith population and subtracting the constant p_0. But subtracting a constant does not affect the variance. Hence

$$V(\delta p_i) = \frac{p_0^* q_0^*}{2N}$$

We shall require $E(\delta p_i)^2$. As a matter of definition

$$V(\delta p_i) = E(\delta p_i)^2 - [E(\delta p_i)]^2$$

so that

$$E(\delta p_i)^2 = \frac{p_0^* q_0^*}{2N} + (\alpha p_0 q_0)^2$$

or about

$$\frac{p_0^* q_0^*}{2N}$$

when α is small.

In attempting to use these results in order to find the probability of survival and fixation of advantageous allele $\underline{A}$, we follow a procedure which may strike the reader as bizarre or even comic. We make whatever approximations seem plausible, in order to get an answer which will be near enough exact under *some* conditions. We then check whether the answer holds under a more general set of conditions. The author is indebted to a former student for the comment "the trouble with this subject is that you guess the answer and then show that your guess is right"—a comment which would be appropriate in many areas of theoretical research. Anyone reading the literature of population genetics will indeed be struck by the remarkable facility for happy approximations displayed by its leading practitioners.

This being so, we shall assume temporarily that in addition to α being small enough for us to neglect terms in $\alpha^2, \alpha^3, \alpha^4, \ldots$, N is large enough for us to neglect terms in α/N so that we have the approximation

$$E(\delta p_i)^2 = \frac{p_0^* q_0^*}{2N} = \frac{(p_0 + \alpha p_0 q_0)(q_0 - \alpha p_0 q_0)}{2N}$$

$$= \frac{p_0 q_0}{2N} \quad \text{near enough}$$

We have shown: if we write δp_i for the change in frequency of $\underline{A}$ in the ith population when going from newly formed zygotes in generation 0 to newly formed zygotes in generation 1, we may use the approximations

$$E(\delta p_i) = \alpha p_0 q_0 \qquad E(\delta p_i)^2 = \frac{p_0 q_0}{2N}$$

at least in cases where α is small and N large. It may be shown, from the standard properties of the binomial distribution, that in such cases

$$E(\delta p_i)^r \quad \text{is negligible}$$

for any r *greater than* 2.

The probability of fixation: Kimura's formula

As we have seen, the probability of fixation depends on the values of α, N and p_0. Suppose we decide on values for α and N but postpone a decision as to the value of p_0. The probability of fixation will depend on our ultimate choice of value for p_0, so that our probability is a function of p_0 which we shall write $u(p_0)$. Thus we have decided on our destination, fixation, but not yet our starting point, rather as if we had decided to go to Ireland by ferry but had not yet made up our minds whether to leave from Pembroke or from Fishguard. A set-up of this kind gives rise to what are known as "backward" equations, whereas "forward" equations arise when the starting point is fixed but not the destination. "Give me a firm spot on which to stand, and I will move the earth"

(Archimedes)—backward equations arise when the firm spot is the destination.

Suppose then we start (generation 0) with some value of p_0, as yet undecided, giving probability of fixation $u(p_0)$. One generation later, the frequency of $\underline{A}$ will usually have changed to some other value; if the ith potential population is actually realized in practice, the new frequency will be $p_0 + \delta p_i$ and the new probability of fixation $u(p_0 + \delta p_i)$. Of course we do not know which potential population is realized in generation 1; we shall have eventually to make our forecast on the basis of the information we have, namely p_0 (when we decide what it is). However, we can say

probability of ultimate fixation given p_0 in generation 0
= Σ[(probability of going from p_0 in generation 0 to $p_0 + \delta p_i$ in generation 1)
 $\times$ (probability of ultimate fixation given that we go to $p_0 + \delta p_i$ in generation 1)]

the sum being taken over all possible values of δp_i. The second probability in the above expression we have agreed to call $u(p_0 + \delta p_i)$. The first probability, being the probability of getting a particular frequency $p_0 + \delta p_i$, must also be the probability of getting the corresponding value of $u(p_0 + \delta p_i)$. In other words, we have multiplied every possible value of $u(p_0 + \delta p_i)$ by the probability of getting it and summed over all δp_i. By the definition of mean, this is just the mean value, over all δp_i, of $u(p_0 + \delta p_i)$. In symbols

$$u(p_0) = \underset{\delta p_i}{E}\, u(p_0 + \delta p_i)$$

Now the allele frequency at any time necessarily changes in discrete steps of $1/(2N)$ but, unless N is small, these steps will be small ones, so that the allele frequency may be treated as if it were a continuous variable. Expanding in Taylor series (appendix 2) we have

$$\underset{\delta p_i}{E}\, u(p_0 + \delta p_i) = \underset{\delta p_i}{E}\, u(p_0) + \underset{\delta p_i}{E}\, \delta p_i \frac{du(p_0)}{dp_0}$$
$$+ \underset{\delta p_i}{E}\, \frac{(\delta p_i)^2}{2!} \frac{d^2 u(p_0)}{dp_0^2} + \underset{\delta p_i}{E}\, \frac{(\delta p_i)^3}{3!} \frac{d^3 u(p_0)}{dp_0^3} + \cdots$$

Now $u(p_0)$ and the derivatives $du(p_0)/dp_0, d^2 u(p_0)/dp_0^2, \ldots$ are not functions of (δp_i) and may therefore be treated as constants when averaging. Writing in

$$\underset{\delta p_i}{E}\, (\delta p_i) = \alpha p_0 q_0, \quad \underset{\delta p_i}{E}\, (\delta p_i)^2 = \frac{p_0 q_0}{2N}, \quad \underset{\delta p_i}{E}\, (\delta p_i)^r = 0 \quad (r > 2)$$

we have, to a good approximation (for small α and large N at least)

$$u(p_0) = \mathop{E}_{\delta p_i} u(p_0 + \delta p_i) = u(p_0) + \alpha p_0 q_0 \frac{du(p_0)}{dp_0} + \frac{p_0 q_0}{4N} \frac{d^2u(p_0)}{dp_0^2}$$

or, writing u as short for $u(p_0)$,

$$\alpha p_0 q_0 \frac{du}{dp_0} + \frac{p_0 q_0}{4N} \frac{d^2u}{dp_0^2} = 0$$

This is a special case of the "Kolmogorov backward equation" which in its more general form is very prominent in the literature dealing with fixation probabilities (Kimura 1964, Maruyama 1977). The approach via the backward equation provides much the easiest method for solving such problems (the reader who has followed us so far and is perhaps inclined to doubt this last statement should look at Fisher's or Wright's treatments).

Cancelling out the $p_0 q_0$ we have the equation

$$\frac{d^2u}{dp_0} + 4N\alpha \frac{du}{dp_0} = 0$$

from which u can be obtained as follows. Integrating, we have

$$\frac{du}{dp_0} + 4N\alpha u = A$$

where A is an integration constant to be evaluated later. To get u, we must carry out a further integration. One way of doing this is to first multiply both sides of our equation by

$$e^{4N\alpha p_0}$$

giving a left-hand side

$$e^{4N\alpha p_0} \frac{du}{dp_0} + 4N\alpha e^{4N\alpha p_0} u = \frac{d}{dp_0} (e^{4N\alpha p_0} u)$$

from the usual rules for the differentiation of a product. Thus

$$\frac{d}{dp_0} (e^{4N\alpha p_0} u) = A e^{4N\alpha p_0}$$

which on integration yields

$$e^{4N\alpha p_0} u = \int A e^{4N\alpha p_0} dp_0 = \frac{A}{4N\alpha} e^{4N\alpha p_0} + B$$

B being another integration constant. For convenience writing $A/(4N\alpha) = C$ (a constant) we have, on multiplying both sides by $e^{-4N\alpha p_0}$

$$u = C + Be^{-4N\alpha p_0}$$

(the general theory of equations involving differential coefficients tells us that all solutions of *our* backward equation are subsumed under this formula; to find the particular one of these solutions relevant to our problem, we must use our understanding of the biology of the situation to evaluate the constants C and B).

Now suppose p_0 were 0, fixation could not occur, that is, if $p_0 = 0$, $u = 0$. Inserting these values, we get

$$0 = C + Be^0 = C + B$$

(any number to the power 0 equals 1), so that $B = -C$, giving

$$u = C(1 - e^{-4N\alpha p_0})$$

If p_0 were 1, fixation has already occurred and is thus certain, that is, if $p_0 = 1$, $u = 1$ giving

$$1 = C(1 - e^{-4N\alpha})$$

or

$$C = \frac{1}{1 - e^{-4N\alpha}}$$

so that we have, at last

$$u(p_0) = \frac{1 - e^{-4N\alpha p_0}}{1 - e^{-4N\alpha}}$$

for the probability of survival and fixation, starting from an initial frequency p_0. We shall refer to this formula for $u(p_0)$ as *Kimura's formula*.

A special case of very great interest is the probability of survival and fixation for a mutant which has just arisen; we have $p_0 = 1/(2N)$, giving us

$$u(p_0) = \frac{1 - e^{-2\alpha}}{1 - e^{-4N\alpha}}$$

Now $e^{-2\alpha} = 1 - 2\alpha + \text{terms in } \alpha^2, \alpha^3, \ldots$ (appendix 2) so that when α is small and we can neglect terms in $\alpha^2, \alpha^3, \ldots$ we have the approximation for the case $p_0 = 1/(2N)$

$$u(p_0) = \frac{2\alpha}{1 - e^{-4N\alpha}}$$

as found by Fisher (1930) and Wright (1931). If $N\alpha$ is also large,

$$e^{-4N\alpha} = 1/e^{4N\alpha} \text{ is close to zero}$$

and Fisher and Wright's formula reduces to 2α, as given by Haldane (1927).

Accuracy of Kimura's formula

Kimura's formula is simple and (see later) easy to interpret. However, in deriving the formula we have made many approximations; nevertheless, we should like to keep to this formula, since it seems that a more exact formula, such as that obtained by Ewens (1964), will be very complicated. Hence we must establish the conditions under which Kimura's formula holds. One way of doing this is to take the formula and substitute for $u(p_0)$ and $u(p_0 + \delta p_i)$ in the (exact) equation

$$u(p_0) = \mathop{E}_{\delta p_i} u(p_0 + \delta p_i)$$

After substitution, the left-hand side (LHS) should equal the right-hand side (RHS) under conditions where Kimura's formula holds. The denominator

$$1 - e^{-4N\alpha}$$

is common to both sides. The numerator on the LHS is, of course,

$$1 - e^{-4N\alpha p_0}$$

while that on the RHS turns out to equal

$$1 - [1 - p_0^*(1 - e^{-2\alpha})]^{2N}$$

Using the *exact* formula for p_0^* it may then be shown that the discrepancy between the numerators of the substituted left and substituted right sides does *not* involve terms in α *or* in α^2 but only in $\alpha^3, \alpha^4, \ldots$. In view of this remarkable fact, agreement should be very good for α small or moderate. Numerical calculations show very close agreement for values of α up to 0.05 and good agreement for α up to 0.10, suggesting that the formula will hold to a high degree of accuracy for most cases encountered in practice.

A different approach is taken by Ewens (1963). For small N, it is possible to obtain exact values for $u(p_0)$. He put $2N = 12$, with p_0

ranging from $\frac{1}{12}$ to 1, and found excellent agreement between results from Kimura's formula and the exact formula for α up to 0.10, the discrepancy never exceeding 3 % of the true value.

Yet another approach is due to Moran (1960). He showed that under all circumstances, the exact $u(p_0)$ will lie between two values, an upper value given by Kimura's formula and a lower value given by taking Kimura's formula and replacing α by $\alpha/(1+\alpha)$. For α up to about 0.01 these lower and upper values are very close to one another and they agree well (within about 6 % of one another) for α up to about 0.05, confirming the fact that Kimura's formula is very accurate for α up to 0.05. For α up to 0.10, upper and lower values do not disagree by more than about 11 % of one another. Thus all the evidence indicates that the formula is very accurate up to $\alpha = 0.05$ and out by only a few percent of the true value up to $\alpha = 0.10$; for larger selective advantages we can at least obtain a rough idea of $u(p_0)$ by calculating the upper and lower values. For example, with $\alpha = 1$ and N anything but miniscule, the chance of fixation of a newly arisen mutant will lie between 0.63 and 0.86. Alternatively, for a mutant *initially rare in a large population*, the probability of fixation may be found from a completely different approach, using the theory of "branching processes" (Fisher 1922, 1930, 1930a; Haldane 1927). It turns out that if we write x for the natural logarithm of $(1+\alpha)$, the probability of fixation of a newly arisen advantageous mutant in a large population is

$$2x - \tfrac{5}{3}x^2 + \tfrac{7}{9}x^3 - \tfrac{131}{540}x^4 + \tfrac{95}{1620}x^5 - \tfrac{771}{68\,040}x^6 + \ldots$$

(other terms will be negligible unless α is very large). For the case $\alpha = 1$, this gives a probability of fixation 0.7967. Details of the procedure are given in e.g. Crow and Kimura (1970).

When N_e (see chapter 3) differs from N (but does not change with time) we just write N_e instead of N in Kimura's formula; p_0, however, is still written in terms of N, so that for a newly arisen mutant we still have $p_0 = 1/(2N)$, *not* $1/(2N_e)$. Varying population size is much more difficult to analyse. To avoid loss of continuity, we postpone discussion of this topic to a later section.

The probability of fixation: some conclusions

In table 4.1, we give the probability of fixation, calculated from Kimura's formula, of a newly arisen mutant, for various values of α and N. To avoid lengthy decimals, we give the probability as a percentage.

Table 4.1 Probability (*per cent*) that a newly arisen mutant will ultimately be fixed (probability given in the body of the table, corresponding to value of α and of N given in the margins)

		α					
		0	0.001	0.005	0.01	0.05	0.1
	100	0.5000	0.606	1.151	2.017	9.516	18.127
	1000	0.0500	0.204	0.995	1.980	9.516	18.127
N	10 000	0.0050	0.200	0.995	1.980	9.516	18.127
	100 000	0.0005	0.200	0.995	1.980	9.516	18.127
	1 000 000	—	0.200	0.995	1.980	9.516	18.127

The first point to notice is that even with α as large as 0.1, the probability of fixation is only about 18%. Indeed, if (for simplicity) we consider a large population, it turns out that the probability of fixation will be less than 50% unless α is as large as 0.39; such a large value must surely be unusual. Hence it is a fallacy to suppose that an advantageous mutant is necessarily fixed. Rather we have

CONCLUSION 1: The overwhelming majority of mutants, even if advantageous, are lost from the population as a result of drift.

We should perhaps stress that this conclusion is *not* a point of controversy.

Our conclusion, however, is not perhaps as important as may appear at first sight, because we are much more interested in explaining what *has* happened in evolution than what has *not* happened. In any population, mutants arise from time to time; even if the chance that any *individual* mutant at a given locus will be fixed is low, mutants at the locus keep appearing in different generations, so that it is a near-certainty that eventually *one* or *other* of them spreads and finally becomes fixed. The point, then, is: which mutants are typical winners? Advantageous mutants or neutral mutants? Suppose, say, that in a given generation one advantageous and 50 neutral mutants arise. The chance of fixation of *one* or *other* of the 50 neutrals taken *en bloc* will just be the chance of fixation of a single neutral allele present initially in 50 copies, i.e. the chance is $50/(2N)$. If the chance of fixation of the advantageous mutant greatly exceeds $50/(2N)$, then *if any fixation of our mutants occurs* at a later stage, it is almost certainly the advantageous mutant that is fixed; the neutrals get nowhere. On the other hand, if the chance of fixation of the advantageous mutant exceeds the chance of fixation of any individual neutral but does not exceed $50/(2N)$ then, if any fixation does occur, one of the neutrals could well be the winner. This simple argument, although helpful, is not altogether satisfactory, since it takes no account of the possibility that the advantageous mutant lost by chance in one generation might reappear *de novo* as a result of mutation in a later generation. We take our problems one step at a time, however, and first consider the extent to which the chance of fixation of a single mutant is enhanced by selective advantage.

For any chosen value of N, read across table 4.1 from left to right. It will be apparent that the probability of fixation rises sharply as α increases. For example, for a population size 1000, the probability of fixation, equal to 0.05 % for a neutral allele, rises to 18.127 % for $\alpha = 0.1$, a 362-fold increase. Even for α as small as 0.005, the increase is 20-fold. Hence

CONCLUSION 2: A selective advantage, even if quite small, substantially increases the probability of fixation of a newly arisen mutant.

Reading the table now row by row, the reader will perceive that, the larger the population size, the more dramatic is the effect of natural selection. For example, on calculating the ratio R (say) defined as (probability of fixation when $\alpha = 0.005$) $\div$ (probability of fixation for a neutral allele) for different values of N, we find

N	R
100	2
1000	20
10 000	199
100 000	1990
1 000 000	19 900

In cases where we may use Haldane's formula, namely 2α, for the probability of fixation, the corresponding R is simply

$$2\alpha \div \frac{1}{2N} = 4N\alpha$$

and this gives a *rough* guide to the true R for our range α 0.001 to 0.1, N 100 upwards, although inaccurate for small $N\alpha$ or large α (the reader can easily check the table for cases where Haldane's formula applies, remembering, however, that for convenience entries in the body of the table are given as percentages). We have, then

CONCLUSION 3: The effect of selective advantage on survival is most marked when the population is large.

Now a persistently very small population is easily wiped out, so that we can probably ignore the possibility that the population size would remain very small for long periods during the evolution of a successful species. Suppose then a moderate population size. Take as example our own species; it has been suggested that the effective population size for much of the period when Man was evolving was about 13 000. Presumably the species at that stage consisted of separate groups which intermated only occasionally but—remarkably—this does not affect the calculation (Maruyama 1977). Write $N_e = 13\,000$ for N in Kimura's formula. The term

$$e^{-4N_e\alpha}$$

will be negligible unless α is minute, giving probability of fixation

$$1 - e^{-4N_e\alpha p_0} = \text{about } 4N_e\alpha p_0$$

for small p_0. For a newly arisen mutant, $p_0 = 1/(2N)$ so that

$$R = 4N_e\alpha\left(\frac{1}{2N}\right) \div \frac{1}{2N} = 4N_e\alpha = 52\,000\alpha$$
$$= 260 \quad \text{for} \quad \alpha = 0.005$$

For many species N_e was presumably much larger. Clearly, then, for every mildly advantageous mutant that arises there must be very many more neutral mutants arising at the same locus if any of the neutrals are to spread. Otherwise, if any fixation occurs, it is the advantageous allele that is fixed, the neutral mutants one and all being lost—"spilled upon the ground" in Lord Keynes's biblical metaphor. Thus

CONCLUSION 4: For evolution at a given locus to proceed according to the neutralist scheme, a necessary condition is that the overwhelming majority of non-harmful mutations at that locus must be neutral or almost neutral.

In chapter 1 we asked "can a selective advantage be so small that it does not matter"? The brief answer "yes" is not very illuminating. Rather we should seek the conditions under which the answer is "yes". Inspection of table 4.1 reveals that for R to be close to 1, $N\alpha$ must be small. How small? The term

$$e^{-4N\alpha} = 1 - 4N\alpha + \text{terms in } (4N\alpha)^2, (4N\alpha)^3, \ldots$$

will be close to $(1 - 4N\alpha)$ when $4N\alpha$ is small; from tables it appears that the approximation is very good if $4N\alpha$ is less than $\frac{1}{4}$ or so. Under these circumstances

$$u(p_0) = \frac{1 - e^{-4N\alpha p_0}}{1 - e^{-4N\alpha}} = \frac{1 - (1 - 4N\alpha p_0)}{1 - (1 - 4N\alpha)} = \frac{4N\alpha p_0}{4N\alpha} = p_0$$

to a very close approximation. Thus R will be near to 1 if (and only if) $4N\alpha < \frac{1}{4}$ or so, that is

$$\alpha < \frac{1}{16N}$$

We write this result as

CONCLUSION 5: Even a small selective advantage will augment the chance of fixation to some extent. Only if the advantage is less than about 1 in $16N$ will the allele behave as though it were neutral.

One point in table 4.1 that may have surprised the reader is that, for given α, the probability of fixation is greatest when the population is small (with α sufficiently large, however, this effect is too small to be apparent in our table). This apparent "paradox" is resolved by noting the obvious point that an advantageous mutant must have chance of fixation at least as great as that of a neutral mutant; the effect of selection being to increase this chance. With rather mild selection, the increased chance of fixation of a neutral mutant in very small populations makes itself apparent, the effect fading out as N increases. More precisely, when $4N\alpha$ exceeds 8 or so, $e^{-4N\alpha}$ is altogether negligible and the probability of fixation is

$$1 - e^{-2\alpha}$$

irrespective of the population size. Only under these conditions is it correct to suppose that the sole effect of drift is the tendency to eliminate the advantageous mutant. We state this as a formal conclusion.

CONCLUSION 6: Provided the population size N and selective advantage α are large enough for $N\alpha$ to exceed 2, the chance of fixation of a newly arisen mutant is the same irrespective of the population size.

Note, however, that in establishing this result we have taken N to be constant over time (see later).

Finally, we consider the case where the same advantageous mutant $\underline{A}$ appears *de novo* several times; r times, say, *not necessarily in the same generation*. What is the probability that *one or other* of these r mutants is eventually fixed? This is clearly a critical question, since in regard to the evolution of the species, it scarcely matters that a specific $\underline{A}$ mutant is lost, provided enough mutations occur for the population to end up all $\underline{AA}$ within a reasonable period of time.

An algebraic solution to this problem is easily found, provided we can assume that the probability of survival of any one of our r mutants is independent of the probability of survival of any of the others, irrespective of the degree of dominance in fitness. This will be the case provided we can assume that the whole matter of survival or loss is settled, for any individual mutant, when the descendants of that mutant are rare. As we shall see a little later, the assumption is justified provided $4N\alpha$ is large (just how large we leave for the moment). We deal only with the case $4N\alpha$ large, so that

probability of fixation of any one of our r $\underline{A}$ mutants $= 1 - e^{-2\alpha}$
probability of loss of any one of our r $\underline{A}$ mutants $= 1 - (1 - e^{-2\alpha}) = e^{-2\alpha}$
probability that all our r $\underline{A}$ mutants are lost $= (e^{-2\alpha})^r = e^{-2\alpha r}$
probability that the population ends up all $\underline{AA} = 1 - e^{-2\alpha r}$

Numerical values are given in table 4.2.

Table 4.2 Probability that the population ends up all $\underline{AA}$ if mutant $\underline{A}$ with selective advantage α arises independently r times

αr	*Probability (in %)*
0.1	18
0.3	45
0.5	63
0.7	75
0.9	83
1.1	89
1.3	93
1.5	95

It will be apparent that, if α is small, the mutant will have to appear a fairly large number of times if there is to be a large chance that the population will end up all $\underline{AA}$. Thus with $\alpha = 0.01$, r must be 150 for this chance to be 95%. For very advantageous alleles, however, quite small values of r will do.

How likely is it that 150 (say) independent mutations will occur? Unfortunately, this question has no simple answer. Consider a particular advantageous "mutant phenotype", that is, a phenotype advantageous owing to some specific defined mode of biochemical action resulting from mutation at a specified locus. In how many ways could a given mutant phenotype arise? At one extreme, the mutant phenotype might represent pure loss of some function. Since changes at many sites within the gene would result in loss of function, the *relevant* mutation rate would be about 10^{-5} to 10^{-6} in such cases. At the other extreme, the mutant phenotype might appear only if a specific base-pair change occurred at a specific site; in that case, the relevant mutation rate could be as low as 10^{-8} or so. If the mutant phenotype were a defined altered temperature or pH optimum for a particular enzyme, the relevant mutation rate would be intermediate between these extremes. Of course, if the population were large enough, the required number of mutants would appear within a relatively small number of generations, even if the relevant mutation rate was very low. We should bear in mind that a very small amount of migration between sub-populations implies that the members of all the sub-populations can be treated as if they were all members of a single population, provided the mutant is about equally advantageous in all sub-populations (we have noted earlier Maruyama's finding that subdivision into sub-populations does not affect the probability of fixation, provided there is some migration between the various sub-populations; a very high degree of isolation will, however, slow down the process of fixation). Thus by the "population" we could mean a substantial part of, or even the whole of, the world population of the species. We have then

CONCLUSION 7: A mutant $\underline{A}$ conveying a small selective advantage will need to recur independently a fair number of times for there to be a good chance of the population ending up all $\underline{AA}$. The necessary number of recurrences would, however, come about in a reasonable number of generations if the population were really large. With large selective advantages, the required number of recurrences would be few.

We have therefore as our final conclusion:

CONCLUSION 8: The selectionist view is most compelling if, in the course of evolution, effective population sizes are typically large. The neutralist view requires that effective population sizes be typically moderate or small.

We shall find other reasons in justification of this last conclusion in chapter 6.

Initial frequency larger than $1/(2N)$

We have seen that if the initial frequency p_0 of the advantageous allele is $1/(2N)$, the allele is usually lost by drift, unless the advantage is very large. If, however, the initial frequency is rather larger than $(1/2N)$, the effect of drift is very much weakened. Of course, if $4N\alpha$ is small (less than $\frac{1}{4}$) so that the allele is effectively neutral, the chance of fixation is still low if p_0 is small. If $4N\alpha$ is large, however (greater than 80, say) things are quite different. Substitution in Kimura's formula makes it clear that if $4N\alpha p_0$ exceeds 8, that is

$$p_0 \text{ exceeds } 2/(N\alpha),$$

fixation is almost certain. An example will make the point clear. Suppose we have an allele with advantage $\alpha = 0.005$ in a population size 13 000. If $p_0 = 1/(2N)$ the chance of fixation is (just less than) 1%, if $p_0 = 2/(N\alpha) = 0.0308$, the chance of fixation exceeds 99.97%. In contrast a neutral allele with initial frequency 0.0308 has chance of fixation 3.08%; thus the advantageous allele has a chance of fixation some 32 times greater in this case (in general, it will be $\frac{1}{2}N\alpha$ times greater). Clearly then if, as a result of the past history of the population, the initial frequency of the advantageous allele, while low, is large enough to exceed $2/(N\alpha)$, the chance of fixation not only greatly exceeds that of a neutral but is very close to unity, and no question of a requirement for further mutation arises.

We may look at the matter in a slightly different way. An advantageous mutation arises, with frequency $1/(2N)$. At this stage, an observer of the population finds the probability of fixation to be about 2α. Generations later, the population is re-examined. It is found that the frequency has happened to rise to about $2/(N\alpha)$, and the probability of fixation calculated on the basis of this happening is found to be near unity. This scenario enables us to give a more exact answer than hitherto to our question "which wins, drift or selection?" As long as the allele frequency remains very low, the fate of the allele is dominated by drift. Indeed, Maruyama (1977) has shown that the probability that the advantageous allele reaches frequency x is hardly greater than the corresponding probability for a neutral allele, provided x is such that $4N\alpha x$ is small (less than $\frac{1}{4}$ or so). Thus it is almost entirely a matter of chance whether the advantageous allele, starting at $p_0 = 1/(2N)$ ever reaches frequency x (in our example, with $\alpha = 0.005$, $N = 13\,000$ we find x to be 0.00096). Given, however, that the frequency has, by chance, reached x, natural selection begins to affect the outcome, both selection and drift being important at this stage. With every rise of allele frequency, selection becomes more and more important vis-à-vis drift in affecting the final outcome. For cases where $N\alpha$ is greater than 2, the effect of drift is altogether negligible in its effect on the final outcome, provided the allele frequency has reached $2/(N\alpha)$. With $N\alpha$ greater than 20, the fate of the advantageous allele is decided for good when the frequency is less than 0.1; this justifies our previous suggestion that under a wide range of circumstances, the whole matter of fixation is settled while the allele is still relatively infrequent.

Variable population size (initial frequency $1/(2N)$)

We have shown that if $N\alpha$ is large, the probability of fixation (for given α) is the same for all N, provided N remains constant. It might be supposed, therefore, that the probability of fixation is unaffected by changes in population size, provided N remains sufficiently large for $N\alpha$ to stay large. This supposition is quite wrong. Changes in population size occurring in the first few generations after the mutant first appears are absolutely critical. To understand this, consider the following intuitive argument. If, given a large population, the chance of fixation of a newly arisen mutant does not depend on the population size, it must be the *number* of $\underline{A}$ mutant alleles rather than the proportion that really matters. When discussing these large populations, then, we should strictly think in terms of allele numbers (although, when the population size is large but constant, the distinction between number and proportion is trivial). Thus instead of saying that the probability of fixation of $\underline{A}$ rises rapidly as the frequency of $\underline{A}$ increases, we should more correctly say that the probability of fixation rises rapidly as the *number* of $\underline{A}$ increases (provided we are considering large populations). This being so, an increase in population size will give an increase in number of $\underline{A}$ alleles, and thus an enhanced probability of fixation as compared with that in a population of constant size.

This topic has been discussed, in a much more rigorous way than we have done, by Kojima and Kelleher (1962) and (with rather different assumptions) by Ewens (1967a, 1969); both treatments use the theory of branching processes and are therefore appropriate for *large* populations. We summarize here the results given by Ewens, whose assumptions are in fact those we made when discussing variable population size in chapter 3.

Suppose the population size first increases and then levels off, taking in successive generations values N, $2N$, $4N$, $8N$, $8N$, $8N$,... for example. Then for $\alpha = 0.1$ the probability of fixation turns out to be

Population size when mutant first appears	Probability of fixation (in %)
N	67.20
$2N$	50.67
$4N$	32.12
$8N$	17.61

(the value 17.61 is a little lower than that given for a population of constant size in table 4.1, the discrepancy representing the slight inaccuracy in Kimura's formula when α is as large as 0.1). The effect is dramatic; a mutant appearing when population number is increasing rapidly has a greatly enhanced probability of fixation. Other values of α give equally striking effects.

When the population size changes in a cyclical manner, we should expect the chance of fixation to be greatest if the mutant first appears when the population size is at its lowest (and therefore about to increase), smallest when the population size is at its greatest (and therefore about to decrease) and intermediate otherwise; all this turns out to be so. Thus for a four-stage cycle N, $2N$, $4N$, $2N$ and $\alpha = 0.1$, Ewens finds fixation probabilities (in %)

$$29.85, \ 16.12, \ 7.99, \ 15.14 \text{ respectively.}$$

Now if, a little naively (but see Chia 1968), we took Kimura's formula and substituted N^*, the harmonic mean of the population sizes during the cycle, for N we should find for the probability of fixation (given $4N^*\alpha$ large)

$$1 - e^{-4N^*\alpha p_0} = \text{about } 4N^*\alpha p_0.$$

If at the stage of the cycle when the mutant first appeared the population size were say N_i we have $p_0 = 1/(2N_i)$ giving probability of fixation

$$2\alpha \frac{N^*}{N_i}$$

It is apparent from the fixation probabilities given above that this simple formula is an oversimplification since, although the population has size $2N$ at two stages of the cycle, fixation probabilities are not quite the same in these two cases. However, Ewens shows that the simple formula holds approximately under some circumstances ($\alpha < 0.05$ and cycles short with population sizes not too different at the different stages). This is helpful, because the larger the population size, the more likely is the mutation to occur in the first place, and this is easily allowed for in cases where the simple formula applies. Note that, since the mutation is most likely to appear when its prospects are worst, it is not true that, on average, the fixation probability is the same as that in a population of constant size. It is, in fact

$$2\alpha \frac{N^*}{\bar{N}}$$

when $\bar{N}$ is the *arithmetic* mean of the population sizes during the cycle. With unequal population sizes, it may be shown that N^* is necessarily always less than $\bar{N}$ (for example, for the cycle N, $2N$, $4N$, $2N$ we find $N^* = 1.78N$, $\bar{N} = 2.25N$, $N^*/\bar{N} = 0.79$). Thus the average probability of fixation is always *lower* in a cyclical system of large population sizes than would be the case in a large population of constant size.

Dominance in fitness

So far in this chapter we have drawn all our conclusions from the haploid model. This model has been used repeatedly in population genetics since its use simplifies the calculations and leads to *relatively* simple solutions which are often quite adequate. However, there are situations in which the haploid model yields a poor representation of the course of events. We should therefore discuss, fairly briefly, possible effects of incomplete or complete dominance on the probability of fixation. It may be as well, however, to reassure the reader from the start that, with very minor modifications, all of our previous conclusions will be found to hold good.

The appropriate formula for $u(p_0)$ the probability of fixation is readily established using the approach given earlier for the haploid model. Given

	AA	Aa	aa
Viability	$1+s$	$1+sh$	1

Kimura (1957) found that

$$u(p_0) = \frac{\displaystyle\int_0^{p_0} e^{-\{4Nshx + 2Ns(1-2h)x^2\}} dx}{\displaystyle\int_0^1 e^{-\{4Nshx + 2Ns(1-2h)x^2\}} dx}$$

provided terms in $s^2, s^3, \ldots$ are negligible. For given s, h and p_0 the value of $u(p_0)$ may be found by numerical integration on a computer. Although the formula is approximate, results can be compared with exact results obtained by another method available only when N is small. It emerges that the formula is very accurate for values of s up to 0.1 (Carr and Nassar 1970). The difficulty is that the formula cannot, in general, be simplified (although in special cases some simplification is possible). Thus the detailed implications of the formula are not readily seen, although the broad outlines are sometimes apparent. However, if $u(p_0)$ is evaluated numerically for a large number of special cases, the important points of principle may be established.

Perhaps the easiest way to proceed is to consider the case $p_0 = 1/(2N)$ and ask: under what circumstances does the value of $u(p_0)$ obtained from this formula (the "diploid" formula) agree well with the value obtained by putting $sh = \alpha$ in Kimura's formula given earlier (the "haploid" formula)? The answer depends critically on the value of Ns. When Ns is sufficiently small (less than 0.1 or so) the allele is effectively neutral, virtually irrespective of the value of h. At this point, the two formulae give the same answer. Our conclusion 5, that even a small selective advantage will augment the chance of fixation, still holds. We just replace the condition for effective neutrality α less than 1 in $16N$ with s less than 1 in $10N$. Of course, if $h = \frac{1}{2}$, these conditions are much the same.

With rather larger values of Ns, the two formulae sometimes disagree. However, from our earlier discussions the formulae should agree well if the chance of fixation becomes near-unity when the frequency of the advantageous allele is still fairly small. How small? This depends rather strikingly on the value of h. If h is very small (i.e. the advantageous allele is almost recessive in fitness) the difference in selective advantage between the mutant homozygote $\underline{AA}$ and the heterozygote is very marked. Hence if only a very few $\underline{AA}$ homozygotes are present, they still make a difference to the outcome. It follows that, if the haploid formula is to hold (in cases where Ns is not very small), then the smaller the value of h, the fewer must be the proportion of $\underline{AA}$ homozygotes or of heterozygotes present at the time, if any, when ultimate fixation of the allele becomes near-certain. Since we have equated α to sh and have agreed that the fate of the allele is settled early in cases where $N\alpha$ is large, we surmise that when h is very small the haploid formula gives a good approximation to the truth in cases where N is large enough for Ns to be very large indeed. Generally, the two formulae agree well in cases where Ns is very small and also when Ns is large, although "large" implies "very large" in cases where h is small. For h greater than about 0.2, however, this difficulty is much less acute. Over the range h 0.2 to 1, the haploid formula turns out to be very accurate for Ns greater than about 90, quite accurate (sometimes very accurate) for Ns greater than 45, and out by at most one third of the true value otherwise (to keep the matter in perspective, note that when $N = 13\,000$ and $s = 0.005$, $Ns = 65$). For h near 0.5 the agreement will, of course, be much better. Note also that when h is small, the probability of fixation, while always greater than that given by the haploid formula, will never be large, save with large N and large s. Thus even with small h the result from the haploid formula, although sometimes inaccurate by a large factor, is still the right sort of answer. Thus conclusions 1 to 4 inclusive, stand unmodified; that is, it is still true that most advantageous mutants are lost by drift, that a selective advantage (even if quite small) substantially increases the chance of fixation, especially in large populations, and that the neutralist scheme requires that nearly all non-harmful mutations be neutral (or almost so). Conclusion 6, that with $N\alpha = Nsh$ sufficiently large the chance of fixation is independent of the population size, still holds. However, for very small h the result is true only when Ns is very large indeed. For h between 0.2 and 1, the result holds at least if Ns exceeds 90. Conclusions 7 and 8 now follow automatically from our findings.

Thus it will be seen that our adherence to the haploid model, which may have struck the reader as somewhat perverse, was after all justified. Conclusions from the haploid model came out very easily, whereas much poring over computer print-outs is required before general patterns can be established from the diploid formula. However, if really accurate results are required in any individual case where we have reason to believe that the two formulae disagree, the diploid formula must of course be used.

This ends our rather lengthy discussion on the probability of fixation of an advantageous mutant, a length we consider justified by the importance of the topic. We conclude with a word of warning. It will take time for a mutant, even a very advantageous mutant, to increase in frequency up to fixation. Of course, we have ignored the possibility of mutation of the advantageous mutant to disadvantageous mutants, but this will not affect the outcome to any very significant extent (it hardly matters whether the mutant ends up fixed or nearly-fixed). More important is our tacit assumption that the mutant, while spreading, retains its advantage over all rival alleles at the locus, in spite of changes in the environment. Thus a long-term prediction, made when the allele frequency is still low, that ultimate fixation is near-certain is subject to the assumption that conditions remain much the same for some time in the future.

> *Clarence still breathes; Edward still lives and reigns;*
> *When they are gone then must I count my gains.*
>
> (Richard III)

Summary

The fate of an advantageous allele depends upon the action of drift as well as that of natural selection. When the advantageous allele is very rare, its future is decided almost entirely by the effect of drift. Hence most newly arisen advantageous mutants fail to become established. However, if by chance an advantageous mutant survives this initial possibility of loss, the effect of natural selection becomes very apparent. Only if the selective advantage is really small will the probability of ultimate fixation be much the same for a newly arisen advantageous allele as for a newly arisen neutral allele. Even if the advantage is quite small, the probability of fixation will be substantially greater for an advantageous allele, particularly when the population is large. Hence the theory that a fair proportion of mutants ultimately fixed are neutral requires that nearly all non-harmful mutants are neutral and that population sizes are typically moderate or small. Otherwise, advantageous mutants will constitute the large majority of mutants actually fixed. With very large population sizes, a given type of advantageous mutant $\underline{A}$ will appear *de novo* many times over various generations. Although the chance of fixation of any particular one of these mutants will usually be small, the chance that the populations ends up all $\underline{AA}$ increases with every fresh mutation to $\underline{A}$, in such a way that a very large population will end up (virtually) all $\underline{AA}$ eventually, unless the selective advantage is very small. If as a result of a change in environment an allele in a large population becomes advantageous at a time when its frequency is low but not very low, its chance of fixation is a near-certainty, again unless the selective advantage is very small. These findings are not affected by the division of the population into sub-populations.

CHAPTER FIVE

WHAT IS THE ROLE OF DRIFT IN EVOLUTION?
I. MORPHOLOGICAL CHARACTERS

"Oh, quite enough to get, sir, as the soldier said ven they ordered him three hundred and fifty lashes", replied Sam.
"You must not tell us what the soldier, or any other man, said, sir", interposed the judge; "It's not evidence".

Charles Dickens, *Pickwick Papers*

The problem

Our discussion so far has been non-controversial, but now, in attempting to assess the role in evolution of random genetic drift, we enter an area of the most profound disagreement. In the view of some population geneticists, drift is of quite minor importance save in special circumstances (for example, when an allele frequency is close to zero). In the view of others, the genetic composition of natural populations is substantially affected by drift, whatever the frequencies of the alleles involved. Now it will be clear from earlier chapters that one's opinion on the possible importance of drift will be intimately bound up with one's opinion on the possible existence of neutral alleles. Suppose it is commonly the case that many alleles at a locus are neutral *inter se*, that is, they have equal effects on fitness. The relative frequencies of such alleles will be purely a matter of chance. Different species would often differ in their allelic composition at a homologous locus, but in many cases this difference would arise through the chance fixation of one allele in one species and a different allele in the other. In a similar manner, *completely isolated* populations of the same species would gradually diverge genetically. Different sub-populations of the same population might or might not diverge, the degree of divergence being governed, in such cases of neutral alleles, by the amount of migration between sub-populations. Polymorphism, in these cases, arises through the chance spread of alleles

72

that happen to have escaped chance extinction and yet have not reached fixation (Kimura and Ohta 1971).

If, however, different alleles at a locus usually differ substantially in their effects on fitness, our interpretation of the genetic composition of natural populations would be quite different from that just given. Species differences would normally imply that a given species has fixed, out of all the alleles available to it (which have managed to survive chance loss when very rare) that allele which leads to the greatest fitness, given the environment of that species and its genetic constitution at loci other than the locus under consideration. Divergence between populations or sub-populations would normally represent different patterns of natural selection in different localities. Polymorphism would arise, in the main, either because a favourable allele was spreading through a population or because natural selection operated in such a way as to maintain genetic variation (see chapter 1; we discuss this last point in detail in chapter 10).

It may indeed be argued that no two alleles will have *exactly* equal effects on fitness. While this may perhaps be so, it is not an effective objection to the neutralist case. If the difference in fitnesses conferred by two different alleles is sufficiently small, although not strictly zero, the fate of the one allele *vis-à-vis* the other will be decided by chance rather than by selection. We saw an example of this in chapter 4 when considering probability of fixation. Thus it is not an abuse of language to describe an allele as "almost neutral" in relation to some other allele. The implication is that, while the alleles may differ by a very little in effect on fitness, we can ignore this difference in calculations and still get the right answer. Of course, the extent to which we can ignore such a difference will depend on the (effective) population size since, as we illustrated in chapter 4, a selective difference negligible when the population size is small may yet have an important effect on the outcome in a large population.

Harmful mutants

It may be as well, before discussing disagreements, to note some points which are *not* controversial. All are agreed, for example, that a substantial proportion of mutants convey a selective disadvantage, at least when homozygous. If we confine our attention to mutations which lead to some change in the amino acid composition of some polypeptide, the point is obvious enough on *a priori* grounds, since a random change in the chemical composition of a protein must often upset the functioning of that protein. In practice, a considerable proportion of the vast number of known mutants give substantially reduced viability or fertility (or both), at least when homozygous, even in cases where the experimenter attempts to raise individuals under conditions where selection is minimal. There can be little doubt that under the much more stringent conditions prevailing in nature such reductions in fitness would be very widespread. However, only qualitative conclusions can be drawn from these findings, because the more drastic the morphological effects of the mutation, the more likely is such a mutation to be noticed and therefore investigated; but the more marked the morphological effects, the more likely (in general) a reduced fitness. As an approach to a more quantitative assessment, we consider the very interesting investigation of Mukai and Cockerham (1977) on spontaneous mutation rates at five loci coding for enzyme structure in *Drosophila melanogaster*. All five enzymes they studied are critical for the life of the fly. For example, flies lacking α-glycerol-3-phosphate dehydrogenase are unable to maintain sustained flight, and flies lacking α-amylase would survive only in conditions where free sugars were available in sufficient quantity. Thus loss of activity of any of these enzymes would be very damaging at least. Of 20 mutants at these loci, detected electrophoretically, only 3 gave a change in banding pattern, representing a mutation to a structurally changed but still active enzyme (at least to some extent) whereas 17 mutants were null, representing loss of all enzyme activity under experimental conditions. A little care is needed in extrapolating to conditions *in vivo*; for example, the loss of activity could arise if an enzyme were active but unstable under experimental conditions, whereas it might be stable under natural conditions. However, such qualifications would be relevant only occasionally; we shall not go far wrong in concluding that (at least) 85% of the mutants detected in this experiment are very harmful. We should, however, bear in mind that the proportion could well be lower for mutants not detectable electrophoretically, and certainly would be if we included nucleotide changes which did not result in any change in the corresponding enzyme.

Obviously adaptive characters

Just as there is general agreement that a fairly large proportion of mutants are harmful, so there is equal agreement that obviously adaptive traits represent the effects of natural selection. It is hardly necessary to state that say, the human eye has reached its present state as a result of the accumulation of mutants at many loci giving increased acuity of visual perception, although it might be the case that minor variations between individuals are of no adaptive significance. Similar considerations apply to those physiological, biochemical or behavioural traits which are clearly adaptive.

"Minor" morphological variants

Sometimes, however, the biological significance of a particular structure is obscure. For example, in the long-headed poppy, hairs on the flowering stalk are appressed, whereas in the closely related field poppy these hairs are usually erect. Occasionally, one finds populations of field poppy polymorphic for this trait, some plants having appressed and some erect hairs, the distinction being under genetic control. Is this all a matter of chance? Or is it the case that individuals of one species are fitter with appressed hairs, individuals of the other species with erect, save in special environments where selection is such as to lead to polymorphism? In this case the answer is unknown. Indeed, investigations aimed at answering such questions are usually very time-consuming, so that the number of cases studied *in detail* is still comparatively small. However, in a perhaps surprisingly large proportion of such studies, individuals differing in respect of apparently "minor" characters turned out to differ very markedly in fitness, a point stressed by Ford (e.g. 1975) and Mayr (e.g. 1963). We must be a little careful in generalizing from these studies. It would not, for example, be fair to include industrial melanism where the very rapid increase in frequency of melanics (implying very intense selection—Haldane 1924) was the main reason why this case was noticed and investigated. The four cases we shall now give are not open to this objection. In all of them, a neutralist explanation has been proposed as a serious possibility, at least tentatively. There seems no reason to suppose, therefore, that the investigators in these cases merely discovered selection that was obvious from the start, and indeed the high intensity of selection actually came as something of a surprise.

Chromosomal inversions in *Drosophila pseudoobscura*

In many populations of *Drosophila*, individuals of the same species differ in respect of their chromosome structure. A study of the banding of salivary gland chromosomes shows that sections of chromosome which go one way in some individuals are inverted in other individuals; individuals heterozygous for the two types of chromosome show the characteristic inversion loops at meiosis. Differences between chromosomes may be quite elaborate, sometimes representing included or overlapping inversions.

This phenomenon, called *inversion polymorphism*, is known in over 30 species of *Drosophila* and in many other genera. The most detailed studies have been made in *Drosophila pseudoobscura* and *Drosophila persimilis*. Both species have five pairs of chromosomes, the rearrangements being usually found in chromosome III; 16 arrangements are known for this chromosome in *D. pseudoobscura* and 11 in *D. persimilis*, of which only one "Standard" is common to both.

We shall discuss some of the work of Dobzhansky and his colleagues on *D. pseudoobscura* (for more detailed discussion and references, see Dobzhansky 1970 or Ford 1975). Consider the arrangements Arrowhead (AR), Standard (ST) and Chiricahua (CH), common in large areas of the United States. The frequency of the different types was found to show a regular cycle of changes during the year, the details varying with locality. Thus at Piñon Flats, Mount San Jacinto, Southern California, the ST chromosome, at frequency 53% in March, fell steadily to a frequency of 28% in June and then rose steadily to 50% in October. At the same locality, the CH chromosome, at 24% in March, rose steadily to 40% in June and then fell steadily to 20% in October. The frequency of AR fluctuated. There was little change over the winter. These studies were conducted from 1939 to 1946, much the same results being obtained every year.

Further north, at Mather, ST increased throughout the season, AR decreased throughout the season, whereas CH remained about constant; the proportions returning to their spring values during hibernation. Such cycles have been detected in many other localities.

We shall discuss possible explanations for these seasonal changes in chapter 10. For the moment, it will be sufficient to note the magnitude of the changes in frequency. *D. pseudoobscura* passes through no more than six to eight generations per year at moderate elevations in California and probably less in most other localities. Breeding populations are too large for drift to have any but the most minute effect over this small number of generations. Clearly, we are dealing here with cases of very intense selection, the relative fitnesses conveyed by the different inversions showing very marked changes with time of year.

Colour and banding in *Cepaea nemoralis*

The snail *Cepaea nemoralis* shows a remarkably versatile polymorphism both for shell colour and banding. The genetics of these characters is summarized in Cain, Sheppard and King (1968); we give only an incomplete summary here, that is, we shall not deal with every detail of the phenotype.

The shell colour is controlled by a series of multiple alleles at a single locus: $\underline{C}^B$ brown, $\underline{C}^{DP}$ dark pink, $\underline{C}^{PP}$ pale pink, $\underline{C}^{FP}$ faint pink, $\underline{C}^{DY}$ dark yellow, $\underline{C}^{PY}$ pale yellow, every such allele being dominant to those succeeding it.

Bands are numbered 1 to 5 starting at the top; 0 is used to indicate the absence of a specific band. For example, 12345 indicates a snail with all five bands present, 00345 indicates a snail with the two upper bands missing. An allele $\underline{B}^0$, unbanded, is dominant to $\underline{B}^B$, banded. At another locus, $\underline{U}^3$, midbanded (00300) is dominant to $\underline{U}^-$, unmodified. At yet another locus, $\underline{T}^{345}$, upper bands suppressed (00345), is dominant to $\underline{T}^-$, unmodified. $\underline{B}^0$ is epistatic to modifiers of banding, that is, an individual carrying $\underline{B}^0$ is unbanded, irrespective of genotype at the $\underline{U}$ or $\underline{T}$ loci. Similarly $\underline{U}^3$ is epistatic to $\underline{T}^{345}$. Finally, the $\underline{C}$ and $\underline{B}$ loci are closely linked, whereas $\underline{U}$, $\underline{T}$ are not linked to $\underline{C}$, $\underline{B}$ or each other. Five other loci are known which also affect bands.

Now the proportions of the various colour and banding types vary with locality.

Consider first the well-known work of Cain and Sheppard (1950, 1954) in the Oxford district. They pointed out that when a snail is seen from above, bands 4 and 5 are hidden by the curvature of the shell and band 3 is only partially visible; from the viewpoint of a predator approaching from above, possible cryptic effects associated with presence or absence of bands would depend almost entirely on presence or absence of bands 1 and 2. They therefore classified all types with 1 and 2 missing as "effectively unbanded". They found that the commonest phenotypes in any habitat were those least conspicuous, to the human eye, against the prevailing background. Thus the proportion of yellow was high in green areas, that of pink high on leaf litter, and that of dark pink and brown in beech woods. The proportion of effectively unbanded was high in a uniform habitat (e.g. dense woodland) and that of effectively banded high in a diversified habitat (e.g. mixed hedgerow).

The chief predator is the Song Thrush, which carries the snail to a nearby stone ("anvil") on which it breaks open the shell. The broken shells accumulate; thus it is possible to determine which type of individual is predated in any given habitat. It turned out that forms inconspicuous to the human eye were also inconspicuous to the predator. Thrushes captured their prey selectively, destroying an unduly large proportion of those whose colour and pattern were ill-matched to the background ("visual selection"). For example, in one case studied by Sheppard, about 1.5 times as many yellows appeared on the anvils as would be expected if predation were random.

We should note that different habitats of the same type were geographically interspersed with habitats of quite different type. The fact that particular morphs were common in one type of habitat rather than another cannot be attributed to migration between adjacent habitats and is explicable only on the assumption of natural selection of some kind. To estimate the intensity of such selection is more difficult. A comparison of frequency of morphs on anvils with the corresponding frequencies among live snails will give the relative selective advantages of the different morphs among those snails actually predated. To translate these into values for all the snails in a given habitat, we need an estimate of the proportion of snails predated. On the basis of such estimates, Cain and Sheppard (1954) arrived at values for selective differences between some pairs of morphs, attributable to thrush predation, of about 1% or more. Probably this figure of 1% is a serious underestimate, since the proportion predated was estimated for only a part of the season. The reader will recall that a selective difference of 1%, although apparently small, is sufficient for selection to overwhelm the effect of drift (given non-extreme allele frequencies) unless the breeding population is very small. There can, moreover, be little doubt that other forms of selection must be operating, apart from that just described, since the presence of a minority of apparently disadvantageous morphs in the different habitats is difficult to explain otherwise. At any rate, it is clear that the selection operating is sufficient to maintain the general correspondence between morph and background.

Visual selection is not, of course, confined to the Oxford district. However, in some places selection appears to be due to climate ("climatic selection"). Consider sand dune populations of C. nemoralis (Cain 1968; Clarke, Diver and Murray 1968). The snails live mainly in regions of marram grass, sometimes admixed with other grasses and herbs. The overall picture suggests visual selection, yellow and pink which (to the human eye at least) give a good match to the background being much commoner than brown. Perhaps glow worms are an important predator

(O'Donald 1968). However, there is much variation, often over short distances within one group of dunes. Most interestingly, these differences can be preserved over a considerable period of time in some cases. The dunes at Bundoran, County Donegal, Eire, were studied by Diver and Boycott in 1924. The same sites (more or less) were revisited by Clarke, Diver and Murray in 1961. Very little change had occurred at the sites in the intervening period. Presumably even snails should have migrated somewhat during this period; it seems likely then that natural selection is responsible for maintaining these differences in the face of migration and intermating. Tentatively, we can conclude that natural selection is operating on snails living on sand dunes, but that some at least of the selection is non-visual. Cain carried out a detailed investigation on one sand dune system (at Point of Air, Clwyd, Wales) and found that brown was associated with areas of complex topography, containing pits or large enclosed areas capable of retaining cold air, and suggested that brown may be at an advantage under such a local climatic condition. Now at Point of Air, very nearly all browns are dark browns, although a lighter form of brown is found on dunes in some other localities. We have then a prediction: the association between dark brown and topography should be found on dunes elsewhere. Twenty-three other dune systems have been surveyed in less detail; this expected association was found at nearly all of them.

Climate has been suggested as an explanation for the frequencies of morphs of *C. nemoralis* in many habitats other than sand dunes, but with a less happy outcome. Predictions often fail (Jones 1973, Harvey 1976). An experimental rather than a purely descriptive approach seems indicated; a comparison of the survival of different morphs from the same locality under different experimental conditions of, say, temperature, humidity and light, repeated for *Cepaea* from many localities, would be helpful. Although such experiments have been attempted, they have been carried out in a rather haphazard manner, the morphs being compared often coming from different localities, as pointed out by Clarke, Arthur, Horsley and Parkin (1978). Obviously, little of use could emerge from comparison of this latter kind; any detected differences in survival could be due to differences between localities at loci quite unrelated to those affecting the appearance of the shell.

Finally, we consider situations where the effects of visual selection are over-ridden by some other factor(s). On the chalk downs of Southern England, there is apparently no correspondence with background. A few morphs predominate over a very wide area, irrespective of local habitat;

sudden "inexplicable" changes in morph frequencies may occur over a very small distance (Cain and Currey 1963). The Marlborough Downs were studied in great detail. In an area of several square kilometres, there were no 5-banded *C. nemoralis*, although these predominated in a contiguous area. Part of the non 5-banded area had a vast excess of brown, another part of yellow. Morph frequencies sometimes changed abruptly over 100–300 metres. There was no correlation between morph and habitat. Patterns of variation of this kind are known as "area effects" (analogous phenomena have been found in other animals).

Goodhart (1963) and Wright (1978) have attempted to account for area effects in terms of the combined action of drift and selection; their explanations, however, differ in detail. Goodhart suggested that if the ancestral populations colonizing the downs in the distant past were small, they might by chance have become adapted to similar environments in quite different ways. Hence when these populations expanded and met, crosses between individuals from different ancestral populations would give progeny combining features of the two co-adapted systems. If the two sets of genes did not work well together, the hybrids would have low viability. Hence there would be a zone of reduced fitness where the populations met, keeping them distinct. Wright would attribute greater importance to drift over the whole period since the ancestral populations became established. He states, "repeated extinction and reestablishment seems probable enough for *Cepaea* colonies, at least over the course of centuries, causing many passages of the line of ancestry of any given colony through many bottlenecks of very small size."

But there is a critical objection to either theory; some area effects are very ancient, going back to well before the Iron Age (about 450 B.C.). It is apparent from the study of sub-fossil snails that area effects have, in general, been rather constant from the beginning of the Iron Age to the present day—the present-day distribution of morphs being virtually established by Roman times (about A.D. 0–200) (Cain and Currey 1968; Cain 1971). This surely disproves Wright's proposal. On the Goodhart theory, we should also not expect the observed constancy, since two co-adapted genotypes would not, in general, be *equally* fit; hence one should gradually replace the other, the boundary gradually shifting at least until a geographical barrier was reached. In fact, one of the more intriguing features of area effects is that they are not, in general, related to such barriers, the same proportion of morphs often being found on opposite sides of a barrier, such as a river (Arnold 1971).

This concludes our very lengthy discussion of colour and banding in *Cepaea nemoralis*. We have thought it worth while to describe the whole problem in detail, in order to indicate the extreme complexity that may be encountered when dealing with natural populations. In the author's view, the dominating role of natural selection for the morphological characters we have described has been clearly established, in spite of this complexity and of our depressing inability to make out many of the details. If the magnitude of selection is not as dramatically large as that found for inversion polymorphisms, it is still large enough to leave little scope for random effects. This conclusion would probably be generally agreed; for a different opinion, see Wright (1978).

Sternopleural chaetae in *Drosophila melanogaster*

We turn now to a metrical character. This case is important, since it has been suggested that while conspicuous characters might be subject to a fair amount of selection, this may not be the case for quantitative variation of a rather inconspicuous type.

Sternopleural chaetae are thought to have some sensory function. Although under artificial selection the number of chaetae may be readily reduced to a mean of about 10 or increased to a mean of about 50, in natural populations the mean number of chaetae is roughly constant from one population to another. In some populations listed by Kearsey and Barnes (1970), for example, the mean ranged from 16.55 to 18.71; variances about these means were fairly small. This rather suggests natural selection in favour of genotypes conferring chaeta numbers within a restricted range, more extreme genotypes being eliminated by natural selection. More tentatively, we might suppose an intermediate optimum, fitness falling progressively with departure from that optimum. Since there is no reason to suppose that the number of chaetae in the adult is particularly critical, we must consider the possibility that the actual natural selection takes place in the pre-adult stage. If this is so, we should have to conclude that the loci controlling chaeta number in the adult are also active in pre-adult life (their function at this latter stage being unknown).

The alternative possibility is that the alleles controlling the number of chaetae are neutral. In that case we should have to attribute the resemblance between populations to migration (a subject we shall discuss in more detail later).

To investigate the possible relationship between chaeta number and fitness, Linney, Barnes and Kearsey (1971) carried out the following experiment. Eggs were collected from a six-year-old cage population ("Texas") and adults raised under conditions of low density, thus keeping competition to a minimum; the adults so raised constitute a control group which have been exposed to a relatively small intensity of natural selection from egg to adult. On the other hand, given the conditions operating in the cage, adults raised in the cage will have experienced high density and hence high competition, and thus a relatively high intensity of selection (at least at some loci). In fact, only about 5–8 % of eggs give rise eventually to adults under such high-density conditions, as compared with about 80 % when the density is low (Barnes, personal communication). Our problem then is: given such a potentially selective regime, does the selection operate differentially (at the pre-adult stage) on loci controlling chaeta number in the adult and, if so, how marked are the differences between genotypes in viability?

Consider any group of genotypes all giving rise to the same mean number of sternopleural chaetae in the adult; call such a group a *genotypic set*. For simplicity of exposition, suppose that under the low-density regime, all such sets are equally viable. Write p_i for the frequency of the ith set among the adults arising from the low-density regime (the values of p_i sum to unity). The corresponding frequency from the high-density regime will be

$$\frac{p_i v_i}{T}$$

where v_i is the (average) viability of the genotypes in the ith set and $T = p_1 v_1 + p_2 v_2 + p_3 v_3 + \dots$. Then

$$\frac{\text{frequency of the } i\text{th set ex the high-density regime}}{\text{frequency of the } i\text{th set ex the low-density regime}}$$

is an estimate of v_i/T. Since only relative viabilities matter here (see chapter 4), we may divide every v_i/T by the largest of them all, say, v_{max}/T, to obtain an estimate of

$$\frac{v_i}{v_{max}}$$

Large differences between sets in v_i/v_{max} indicate large departures from neutrality. This will be true even if our assumption of equal viabilities under the low-density regime, which may not be quite true in practice, does not hold. If u_i is the (average) viability of the ith set under low density, our procedure estimates

$$\frac{v_i}{u_i} \bigg/ \left(\frac{v_i}{u_i}\right)_{max}$$

Thus significant differences between estimates for different sets necessarily indicate selection, irrespective of whether we assume some selection at the low density or not.

There is, however, a complication which may be overcome by appropriate experimental design. Sternopleural chaeta number depends on the environment as well as on the genotype; a given phenotype, say 16 chaetae, on adults from the high-density regime might represent a completely different set of genotypes from that corresponding to the same phenotype on the low-density regime. Linney *et al.* therefore mated a random sample of males resulting from each regime to females of the inbred line Oregon; all progeny were raised under the *same* conditions (in a single randomized block). Any male parent is then characterized by the *mean* chaeta number of his *progeny* which substitutes for his own value in the calculations given above. The mean of such progeny means turned out not to vary significantly with the regime under which the

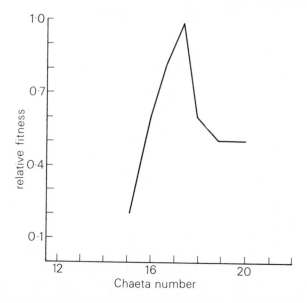

Figure 5.1. The relationship between chaeta number and relative fitness for males sampled from the "Texas" population (reproduced, with permission, from Linney *et al.*, 1971).

male parent was raised, but the variance of progeny means was significantly lower for parents from the high-density regime. This shows that natural selection has operated, genotypes conferring intermediate phenotypes being favoured at high density.

To obtain some notion of the intensity of selection, progeny means were rounded off to the nearest integer, and all parents with the same (rounded off) progeny means assigned to the same set—the relative fitness of that set being then calculated as explained earlier. From the regression of high-density male parents on to their progeny means, the progeny mean for a set can be converted into an estimated parental phenotypic value for that set. Results are given in figure 5.1.

The reduction in fitness with departure from the optimum is quite remarkable. We should note, however, that these large differences in fitness appeared under conditions of intense competition; they are not found when competition is mild (Robertson 1966).

Wing colour in *Panaxia dominula*

In the Scarlet Tiger Moth *Panaxia dominula*, a rare allele $\underline{M}$, when in homozygous form, gives rise to a phenotype known as *bimacula*, in which the black areas on the wings are very much more extensive than in the normal form *dominula* (genotype $\underline{DD}$). The heterozygote, called *medionigra*, although rather variable, is readily identified; it resembles *dominula* more than *bimacula*, but misclassification is unusual (Ford 1975).

In a natural population at Cothill (near Oxford) the $\underline{M}$ allele was very rare before 1929; in museum and other collections (in which rarities are regularly over-represented) Ford found the frequency of the $\underline{M}$ allele to be 1.2%. During the 1930s it became apparent to Ford that the frequency of allele $\underline{M}$ was increasing. Random sampling of the population was begun in 1939 and continued annually (Fisher and Ford 1947, Ford and Sheppard 1969, Ford 1975); the last reference gives sample allele frequencies for every year from 1939 to 1972 inclusive, that is, every generation from 1939 to 1972 inclusive, since this species has one generation per year. The sample frequency of $\underline{M}$ was 9.2% in 1939 and 11.1% in 1940, but since then has declined, with some fluctuations, to 0.8% in 1972. Are these remarkable changes due to drift, as tentatively suggested as a possibility by Wright (1948)? Or must natural selection be invoked to explain them, and if so what has been the intensity of selection operating?

Now the magnitude of changes brought about by drift will, of course, depend on the effective population size. Estimates of the actual population size, based on capture–recapture experiments, were made from 1941 onwards. Effective population size may well have been smaller than the actual size and it is the uncertainty over this, coupled with the absence of data on the population size at the earlier period, which has generated the controversy over these results. It is probably best that the present author should make clear from the start his conviction (which he will attempt to justify shortly) that natural selection has been the major factor in bringing about all the changes in the frequency of the $\underline{M}$ allele. The reader who would like to make a completely independent decision on the matter should consult both Ford (1975) and Wright (1978).

Fortunately, one major point is agreed. The continued decline in frequency of $\underline{M}$ over the period 1940–1972 is, given the population sizes actually found, explicable only on the basis of natural selection; if we ignore the fluctuations over this period for the moment, it turns out that the selective advantage α of the $\underline{D}$ allele over the $\underline{M}$ allele was about 11 %. This settles our main point, that the intensity of selection operating on a polymorphic locus can turn out to be surprisingly large. The initial rise, however, is rather another matter because quantitative estimates of population sizes are not available for the early period. If the rise was due to selection, the implication is an advantage of 20 % or so of $\underline{M}$ over $\underline{D}$. To explain the rise on the basis of drift, we have to suppose an exceptionally small (effective) population size for one or more years, in order to obtain such a large change over such a small number of generations. The problem becomes particularly acute if we are not prepared to accept large changes in relative selective advantage, since if so we must suppose $\underline{M}$ to have been at a disadvantage to $\underline{D}$ during the early period also, so that any increase due to drift would be opposed by selection. Wright suggests near-extinction of the population in one year, with at least half of the population of the next year derived from a single mating in which one of the parents happened to be a heterozygote. There is, however, a serious objection to this view. Ford made observations at Cothill in 1938 and he was able to contact entomologists who had visited the site for all the remaining years in the period 1929–1937 inclusive. A drastic reduction in population size would surely have been noticed. There seems, then, no escape from the remarkable conclusion that the $\underline{M}$ allele switched rapidly from being very advantageous to being very disadvantageous. In the light of this, changes in selective advantage in the later period seem very possible, and an explanation of the fluctuations at this stage in terms of fluctuating advantages very plausible. If, however, the effective population size is substantially less than the actual, a significant contribution from drift at this stage cannot be excluded.

Finally, we should not necessarily suppose that the natural selection is acting directly on the D/M locus rather than on a closely linked locus. We return to this subject in chapter 7.

Conclusions on morphological characters

As a result of the studies we have described and of others given in the references cited, it has become apparent that the intensity of selection acting on morphological characters is often large. It has become generally (but not universally) accepted that while the intensity of selection on such characters may not be as large in general as for some of the cases described, it is still sufficiently intense to overwhelm the effects of drift in the large majority of cases. Of course, this is not an immutable article of faith but a *working hypothesis*. The selectionist may believe that a particular "minor" morphological difference he is studying is of biological importance, but he cannot free himself from the obligation to prove this and to show the manner in which selection acts. Needless to say, this point is so obvious as to be near-universally recognized and would not be worth mentioning were it not repeatedly missed by those who criticize Darwinism on "philosophical" rather than scientific grounds (e.g. some of the contributors to the symposium edited by Moorhead and Kaplan 1967, but the muddle is recurrent). Even the briefest glance at the literature should convince the reader that selectionists like studying selection, and are often prepared to spend many years in the attempt to prove selection and study its working in a particular case. Those interested in the status of Darwinism as a scientific theory should read Maynard Smith (1969); for a brief, interesting account of how the subject might be presented to schoolchildren, see Dowdeswell (1978).

Finally, we should stress the importance of the "ecological genetic" approach. It is the author's firm conviction that studies carried out in the natural habitat, or under experimental conditions simulating leading features of the natural habitat, are essential if we are to make serious progress in our understanding of the factors affecting the genetic composition of natural populations.

(For a summary, see the end of the next chapter.)

CHAPTER SIX

WHAT IS THE ROLE OF DRIFT IN EVOLUTION?
II. PROTEIN VARIATION

Protein polymorphism

By the mid 1960s it seemed that the controversy over the role of drift had largely abated. The view of Wright, that at a large proportion of loci the effects of natural selection and drift are roughly comparable in magnitude, was tentatively rejected by most population geneticists in favour of the view of Fisher and Ford, who argued that the effects of natural selection would dominate those of drift under most circumstances. This, then, was the age of "naive pan-selectionism" (Kimura and Ohta 1971) or "hyperselectionism" (Dobzhansky 1970). We have tried in the preceding chapter to explain how this attitude came about as a result of studies on morphological characters, and have indeed argued that in respect of such characters the strongly selectionist view best meets the evidence available.

However, as Bondi has remarked, "it is certainly a matter of experience that every time our experimental technique has taken a leap forward, we have found things totally unexpected and wholly unimagined before". The advent of techniques, especially gel electrophoresis, enabling us to recognize differences between individuals in the chemical nature of specific proteins (in practice mainly enzymes) has led to the discovery that natural populations are polymorphic to a truly astonishing extent. In *Drosophila*, for example, 53% of such loci are known to be polymorphic (taking an average of the values in different species of this genus) and, although in many cases the percentage is lower than this (being on the whole lower in vertebrates than in invertebrates—see e.g. Selander 1976), it is still remarkably large (e.g. 23% in Man—Harris and Hopkinson 1978). Moreover, these figures are certainly underestimates, since not all differences are detectable by electrophoresis. We discuss this in more detail in chapter 10. For the moment, it will be sufficient to note that while the existence of *some* genetical variation of this kind had long been realized, the actual magnitude, first reported by Harris (1966) in Man and by Lewontin and Hubby (1966) in *Drosophila pseudoobscura* came as a very great surprise. For a historical account of attempts to measure genetical variation within populations, see Lewontin (1974).

With the publication of these results, the controversy over neutral (or almost neutral) alleles and drift began all over again, the universe of discourse now being protein polymorphism rather than polymorphism for morphological characters. The distinction between these two types of polymorphism may seem at first sight artificial, since the morphological differences must ultimately be chemical in nature. But this objection is not necessarily valid, since (with certain exceptions, such as sickle-cell versus normal haemoglobin) there is no conclusive evidence that the detected differences between commonly occurring different forms of the same protein make any difference to the functioning of the organism concerned. It has indeed been shown by Harris and his colleagues that these different forms often differ in respect to important biochemical properties, such as thermostability or affinity for substrates or inhibitors (Harris 1975). Unfortunately, this is not really conclusive, since it is uncertain whether such differences, detected *in vitro*, would necessarily appear *in vivo* and, if they did, whether they would affect fitness. Of course, many of the possible mutants at such loci would certainly be harmful, but these are normally eliminated when still rare. The controversy has centred on those alleles which have reached an appreciable frequency. It is not, of course, denied that some of these will be advantageous, at least under some circumstances. The neutralist contention (Kimura 1968) is that a substantial proportion of them are neutral or almost so, drift alone being responsible for their spread. The same would apply to a fair proportion of the differences between species for a given protein. Indeed, Kimura and Ohta (1971) have argued that most protein polymorphisms constitute a phase of molecular evolution; they state: "those mutants that are destined to spread to the species take a long time until fixation and on their way take the form of protein polymorphism". In a slight variant of the theory (Ohta 1974), which we discuss later, it is supposed that this spread may in some circumstances be counteracted by very mild selection against some almost neutral mutants. Insofar, however, as mutants giving rise to an altered protein do spread, this is attributed to drift in most (but not, of course, all) cases. We should stress that the differences under discussion are differences at loci determining protein *structure*; almost nothing is known on the magnitude of variation at regulatory loci. Note that, while both Wright and Kimura have argued the importance of drift, their theories are quite different. Wright, as we have stated, considers that both drift and selection need to be taken into account when considering changes at most loci. Kimura's protein polymorphism–molecular evolution theory is much more strictly neutralist. In respect of minor morphological characters, however, his position, if the present author understands him correctly, is selectionist.

Rate of nucleotide substitution: theory

An important consequence of the neutral theory is that the rate at which a gene changes in nucleotide composition should equal the mutation rate.

Let u be the "neutral mutation rate" per generation, that is, the probability that a neutral mutation occurs at a given locus in any generation. In a population size N, with $2N$ alleles at the locus, $2Nu$ new neutral mutations arise every generation. As we showed in chapter 3, any one of these has a probability $1/(2N)$ of being ultimately fixed. Let k, the rate of mutant substitution in evolution, be defined as the long-term average of the number of mutants that are substituted in the population per generation. We have the simple result (Kimura 1968)

$$k = 2Nu \times \frac{1}{2N} = u$$

We can if we wish work in terms of years rather than generations. In that case we define u as the neutral mutation rate per year and k as the average number of mutants substituted per year. With these revised definitions k is still equal to u.

On the other hand, a mutant with selective advantage α has probability of fixation

$$\frac{1 - e^{-4N_e \alpha p_0}}{1 - e^{-4N_e \alpha}}$$

where $p_0 = 1/(2N)$ and N_e is the effective population size, as we showed in chapter 4. Clearly, k will now depend in a fairly complicated manner on such variable quantities as the mutation rate to advantageous alleles, the selective advantages of such alleles, and the effective and actual population sizes.

In practice, the large majority of successful mutant substitutions comprise replacement of one base pair by another. To a close approximation, then, k is the rate of nucleotide substitution.

These results look very promising, appearing to provide a clear-cut test of the neutral theory. Assuming for the moment a roughly fixed proportion of nucleotide sites within a given cistron at which sub-stitution does not affect the functioning of the corresponding protein, u and hence k should, *for any given locus*, be constant, if the neutral theory is correct (u and hence k could well vary with locus). On the view that most successful substitutions convey a selective advantage, a constant value of k, at virtually any locus, seems out of the question. The striking contrast between the predictions from the different theories suggests that a final decision between them could be made provided we can obtain satisfactory estimates of k at different stages of the evolution of a representative sample of proteins. Unfortunately, things are not so simple in practice, as we now show.

Rate of nucleotide substitution in practice

We wish to compare the rate of nucleotide substitution (at a given locus) in different lines of descent. Since generation lengths, even on average, will vary from one line of descent to another, we must decide whether we wish to compare rate per year or rate per generation. In practice, as we shall see, it is the rate of evolution *per year* (and not per generation) which is (roughly) constant.

The notion that a given protein evolves at a constant rate over long periods of evolutionary time was first proposed by Zuckerkandl and Pauling (1962). This idea was at first received with much scepticism. Since then a very extensive literature has grown up on this subject. For those wishing to pursue the topic in more detail than will be possible here, we recommend: for an account requiring no previous knowledge, Nei (1975); for a brief, critical look at the procedures used, Fitch (1976); and for a fairly lengthy but readable recent review, Wilson, Carlson and White (1977).

To give some idea of what is involved, we consider first an early analysis by Kimura (1969), after which we briefly review the present state of knowledge.

Consider haemoglobin, say. Take two species, e.g. Man and Carp, and compare polypeptide chains, say α_{man}, α_{carp}. Let n be the total number of amino acid sites in a chain, excluding occasional sites where deletions or insertions have occurred, and d be the number of amino acid sites at which the two chains differ. We have to convert this result into an estimate of the mean number of substitutions per nucleotide site over the whole evolutionary period since the two species diverged from their common ancestor. There are several ways of doing this; in principle, some are better than others, but as in practice all give much the same answer (Nei and Chakraborty 1976) we shall use the method of Zuckerkandl and Pauling (1965) which has the advantage of simplicity. Note that the earlier workers discussed the problem in terms of amino acids rather than nucleotides, and calculated the rate of amino acid rather than nucleotide substitution. Study of the genetic code, however, suggests that we can obtain the total number of nucleotide substitutions from the total number of amino acid substitutions by multiplying the latter by a constant, value 1.3 to 1.4, and this seems to work well in practice (Nei and Chakraborty 1976). Thus we shall get much the same result whether we work in terms of amino acids or of nucleotides. To facilitate reference to Kimura's paper, we shall work in terms of amino acid substitutions.

Let K be the mean number of substitutions per amino-acid site over the whole evolutionary period (since the two species diverged) for both species taken together, so that the total number of substitutions in both species $= nK$. Assume that at most sites substitutions are neutral, with equal probability of substitution at any such site (this latter assumption is not strictly true, but this does not seem to affect the result in any serious way). The probability that a given site remains unchanged in either species on a given occasion when an ultimately successful mutation occurs is

$$\frac{n-1}{n} = 1 - \frac{1}{n}$$

Then the probability that a given site remains unchanged in either species since their divergence from a common ancestor (i.e. the expected proportion of sites unchanged in either species) is

$$\left(1 - \frac{1}{n}\right)^{nK}$$

which is about e^{-K} since n is large (note that we have in fact derived the first term of the Poission distribution). Thus the expected proportion of sites differing in the two species is

$$1 - e^{-K}$$

Provided we are prepared to ignore the remote possibility of parallel substitution at the same site in the two species, or of substitution followed by resubstitution of the original amino acid at a given site, we may equate the observed proportion of sites differing, d/n, to the expected in order to obtain an estimate of K, namely $\hat{K}$. We have

$$\frac{d}{n} = 1 - e^{-\hat{K}}$$

whence

$$\hat{K} = -\log_e\left(1 - \frac{d}{n}\right)$$

Finally, if $\hat{T}$ is the estimated time in years since the two species diverged from their common ancestor, our estimate of $\hat{k}$, the mean number of amino acid substitutions per amino acid site per year in a single line of descent is

$$\hat{k} = \hat{K}/(2\hat{T})$$

Some of the results given in Kimura's paper are quoted in table 6.1. The values of $\hat{k}$ are remarkably alike, particularly since the values of $\hat{T}$ used are unlikely to be a completely accurate reflection of the true times.

Table 6.1 Estimates $(\hat{k})$ of mean number of substitutions per amino acid site per year multiplied by 10^{10} to avoid lengthy decimals, in the evolution of haemoglobin.

Source of estimate	$\hat{k} \times 10^{10}$
Human α, Carp α	8.9
Human α, other mammal αs	8.8
Human α, Human β	8.9
Human β, other mammal βs	11.9
Mouse β, other mammal βs	14.0

Do findings of this kind apply to proteins in general? The answer is a qualified yes. Some of the exceptions may arise from inaccurate dating of divergence times using fossil data. However, Langley and Fitch (1974) have shown how this uncertainty can be overcome; times can be estimated from amino acid sequence data, the unit of time being the average time required for one nucleotide substitution for the proteins and lines of descent under study. In an extensive investigation using this method, Fitch and Langley (1976) demonstrated that, in general, the value of k for a given protein is not exactly constant; the values show statistically significant fluctuations over time, but without any definite trend. A possible explanation is that neutral mutation rates have fluctuated, the proportion of sites at which substitutions do not affect fitness varying in the course of evolution. Nevertheless, values are not so far from constancy that we can forget the whole matter; the rough constancy must be explained. A selectionist explanation, in terms of a more or less constant challenge to species leading to rough constancy of evolutionary rates (Van Valen 1974) has to face the difficulty that rates of evolution for proteins seem unrelated to rates for morphological characters. In the same span of time, one group (e.g. mammals) can show much morphological change, another group (e.g. frogs) remain little changed morphologically, yet both groups show about the same rate of protein evolution. Of course, these results make sense if the morphology is directed mainly by natural selection, whereas the protein changes arise mainly by drift.

The data on the rate of evolution, then, appear to provide a powerful argument for the neutralist view. However, the neutralist explanation is tenable only if the neutral mutation rate *per year*, for a given locus, is about the same in different species, since it is the rate of evolution per year which is roughly constant. Data are too scanty at the moment for any final decision to be made on this point; unfortunately it takes much effort to collect such data. The same applies to a slightly different test of the neutral theory as follows. Different proteins evolve at very different rates. The neutralist explanation is that the neutral mutation rate varies strikingly from one locus to another, mainly because the proportion of sites at which substitutions are neutral is supposed to be very different from one locus to another. Thus protein function should be very much more readily disrupted by amino acid substitutions in a slow evolving protein than in a rapidly evolving one. A laboratory test along these lines may be practicable (Wilson *et al.* 1977).

A different approach

Another approach to the neutralist-selectionist problem is the study of genotype frequencies in present-day populations. We concluded earlier (provisionally) that in cases where genetic differences lead to differences in morphology, genotype frequencies reflect, almost entirely, the action of natural selection. With regard to protein variants of unknown effects, if any, on fitness, the situation is uncertain. Many attempts have been made to discriminate between the neutral and selection theories on the basis of frequency data. We confine ourselves here to discussing just a few points, which should however serve to bring out the difficulties common to all these attempts.

Allele frequencies in geographically distant areas

Consider a number of widely separated communities of a given species. Suppose there are a number of neutral, genetically determined, variants of a given protein. Up to about 1970, it was generally believed that the frequencies of such variants would come by chance to differ markedly from one community to another; indeed, it was thought that in many cases variants present in some communities would be completely absent from others. Clearly, if communities are *completely* isolated, the effects of mutation and drift would be completely independent in the different communities, so that divergence between them in respect of neutral variants would be inevitable. It was thought that this would still obtain, even if isolation was incomplete, provided migration between communities was small. When therefore it was discovered (Stone, Wheeler, Johnson and Kojima 1968; O'Brien and MacIntyre 1969; Prakash, Lewontin and Hubby 1969) that frequencies of such variants (identified by gel electrophoresis) are often very similar from one community to another, even when the communities are a very long way apart, it was concluded that the frequencies reflected the action of natural selection, supposed to act in a similar manner in the different communities.

Effect of migration

The neutralist must invoke migration between communities in order to explain why the allele frequencies are so often similar ("And it shall come to pass in that day, that the Lord shall hiss for the fly that is in the uttermost part of the rivers of Egypt, and for the bee that is in the land of Assyria". Isaiah vii. 18). Since individuals may not as a rule migrate very far, it is necessary to show that quite a small amount of migration between adjacent communities will (in the absence of selection) be sufficient to prevent genetic divergence.

That this might perhaps be so had long been apparent on theoretical grounds. It was indeed clear that a little migration would counteract divergence to a considerable extent; however, it was often thought, on "commonsense" rather than on mathematical grounds, that a fair amount of divergence would still occur. To make a start on the theoretical analysis of the effects of migration, a subject of considerable difficulty, the early writers, Wright and Malécot, had perforce to deal with special cases and the relevance of their results was uncertain. The situation was transformed by theoretical studies carried out in the early 1970s. These studies, mainly by Maruyama (especially 1970, 1970a), while owing much to the work of the earlier investigators, introduced mathematical methods by means of which the difficulties of dealing with more appropriate models of migration were overcome. Unfortunately, the subject is difficult; we shall have to content ourselves here with a few intuitive ideas and leading results. For a review and extensive bibliography see Felsenstein (1976).

As usual, when dealing with a set of communities between which some migration occurs, we shall refer to a community as a "sub-population" and use the term "population" to refer to the totality of sub-populations. Two main models have been discussed, *island* models and *stepping-stone* models. In the former (introduced by Wright), immigrants into any sub-population are a random sample of the whole population. In contrast, in stepping-stone models (introduced by Kimura), immigrants into any sub-population come only from sub-populations immediately adjacent. In principle, the sub-populations in the stepping-stone model might be arranged in an infinitude of possible ways, but it turns out that the critical distinction is whether the sub-populations are arranged in one dimension (along a line or circle) or in two dimensions. Note that it is not really critical to discuss (except in very broad terms) the extent to which these models are realistic; the important point is that the various models span the range of possibilities in respect of migration, and it seems very unlikely that the conclusions obtained will be seriously upset by further refinements. Of course, one could consider some combination of different models, with, say, most migration local but occasional migration from the whole population, and this has been done sometimes. There are also continuous analogues of the one- and two-dimension stepping-stone models; we shall not discuss these here.

Now we must not suppose *a priori* that the effect of migration on divergence will be the same irrespective of the genetical set-up. To investigate whether a particular degree of migration counteracts tendencies to divergence, we have first to define (in broad terms) an appropriate "test situation". Although often such different situations lead to essentially the same conclusions, this is not always so. The set-up most relevant here is the following. We imagine that a large number of different neutral mutants may arise at the locus under study; the number is so large that we may discount the possibility that any newly-arisen mutant was already present in the population before the mutation occurred. We then ask: if we were to examine the population at a long period of time after the pattern of migration was established, would allele frequencies be about the same in the different sub-populations?

Consider first, with Maynard Smith (1970), the island model. Suppose there are r sub-populations, that u is the neutral mutation rate *per generation*, and that m is the probability that an individual breeding in one sub-population was born in another. For simplicity, we may suppose that all the sub-populations start off genetically identical, although this is not at all essential to the argument. Of the neutral mutations that occur, virtually all are lost by drift, but sooner or later one, $\underline{x}$ say, gets going in one sub-population and attains a moderately high frequency there. Now consider the input of *new* genetic material into any other sub-population. Unless the migration rate is very low and/or the number of sub-populations very large, the fraction of the input which is new will consist almost entirely of allele $\underline{x}$, the input of other new mutants, whether arising locally or elsewhere, being trivial in comparison. But, as we saw in chapter 3, the larger the initial frequency, the greater the chance that a mutant will get going. Hence if any new neutral allele does spread, it will almost certainly be allele $\underline{x}$ in all sub-populations. In a rigorous treatment, Maynard Smith shows that

provided m greatly exceeds ur

the sub-populations will all be much alike in the long run or, to put it slightly differently, we can under these circumstances treat all individuals as though they belonged to a single random mating population, of size equal to that of the whole population. Since u is very small (perhaps 10^{-6} or less) it is apparent that a very small amount of migration is sufficient to prevent divergence (save in the case r very large, which is of course very artificial for the island model).

For the stepping-stone model, it will take many generations for $\underline{x}$ to arrive at a sub-population remote from the sub-population in which $\underline{x}$ first got going. But, as Kimura and Maruyama (1971) have pointed out, divergence will not occur if a successful new mutant, such as $\underline{x}$, is able to reach all sub-populations before the *next* successful new mutant has got going. Now the probability of a successful neutral mutant arising in any generation is u. It follows that the average time between successful mutants arising is $1/u$ generations. Thus plenty of time is available for $\underline{x}$ to reach all sub-populations before the next mutant gets going, unless the migration rate is very low indeed. Consider a two-dimensional model arranged in such a way that each sub-population is in contact with four others (to avoid edge effects, the sub-populations can be imagined to lie on the surface of a torus, although a rectangular arrangement gives much the same result). Let m be the proportion of individuals that leave a sub-population in any generation, these emigrants going equally to each of the four neighbouring sub-populations. Suppose there are k sub-populations in all, each size N, so that the total population size $N_T = kN$. Numerical results given by Kimura and Maruyama (1971) derived from the theoretical treatment of Maruyama (1970a) show that for $4N_T u$ up to 0.1, divergence between sub-populations is slight if

$$mN \text{ exceeds } 4$$

For larger $N_T u$, conditions are rather more stringent. When $N_T u$ much exceeds unity, the condition for little divergence is

$$mN \text{ exceeds } 4N_T u$$

Since there are many more routes from one sub-population to another in a two dimensional set-up than in a one dimensional one, we should expect conditions to be more stringent in the latter case, and this turns out to be so. Even here, however, sub-populations may well end up much alike, even with m small, provided k is not too large. In general, of course, the two-dimensional model is much the more appropriate representation of reality, particularly when the number of sub-populations is large. Clearly then, no decision on the relative merits of the selectionist and neutralist views can be drawn from the frequent failure of sub-populations to diverge.

Frequency of heterozygotes

Suppose that we may treat all individuals in the sub-populations as if they were members of just one random mating population. What would be the frequency of heterozygotes?

As before, suppose that any newly appearing neutral mutant is not already present in the population. Consider one locus. Let F_t be the probability that an individual chosen at random in some generation (t) is homozygous. As we have random mating (including the odd chance selfing), our probability is equally the probability that two alleles, randomly chosen from the gene pool, are identical in generation t. Let u be the neutral mutation rate *per generation* and N_e the effective population size.

In the next generation $(t+1)$ two randomly chosen alleles are identical if: (*a*) they are both copies of the same allele in generation t and have not mutated; the probability of this is

$$\frac{1}{2N_e}(1-u)^2$$

or (*b*) they are copies of different but identical alleles in t and have not mutated; the probability of this is

$$\left(1 - \frac{1}{2N_e}\right)F_t(1-u)^2$$

Then the probability that an individual is homozygous in $(t+1)$ is

$$F_{t+1} = \frac{1}{2N_e}(1-u)^2 + \left(1 - \frac{1}{2N_e}\right)F_t(1-u)^2$$

Multiply both sides of this equation by $2N_e(1-u)^{-2}/(2N_e-1)$ and use the approximation (appendix 2) $(1-u)^{-2} = 1 + 2u$, which will be very accurate since u is very small. Then

$$\frac{2N_e(1+2u)}{2N_e-1}F_{t+1} = \frac{1}{2N_e-1} + F_t$$

Let $F = 1/(4N_e u + 1)$ and subtract $2N_e(1+2u)F/(2N_e-1)$ from both sides, giving

$$\frac{2N_e(1+2u)}{2N_e-1}(F_{t+1}-F) = (F_t - F)$$

or

$$(F_{t+1}-F) = \left(1 - \frac{1}{2N_e}\right)\left(\frac{1}{1+2u}\right)(F_t - F)$$

If, then, F_t happened to equal F, the probability that an individual is homozygous would remain at F thereafter. If F_t did not equal F, our result shows that F_{t+1} is a little closer to F than was F_t. Thus the probability that an individual is homozygous, or in other words the expected frequency of homozygotes, at the locus concerned, gradually approaches F over successive generations, although the approach is exceedingly slow if N_e is large. The expected frequency of heterozygotes, then, approaches $1 - F$; writing H for this ultimate expected heterozygote frequency we have (Malécot 1948; Kimura and Crow 1964)

$$H = \frac{4N_e u}{4N_e u + 1}$$

(in Malécot's set-up selfing is excluded; this makes the proof a little more difficult, but yields the same formula for H, within the limits of the approximations used).

At any locus, the actual frequency will usually, by chance, deviate from this value, but we can obtain a value $\bar{H}$ close to expectation by averaging over a large number of loci. The corresponding expected value is obtained by averaging the $4N_e u/(4N_e u + 1)$ over loci; in cases where the $N_e u$ are all small or all very large, this average expected value will be close to

$$\frac{4N_e \bar{u}}{4N_e \bar{u} + 1}$$

where $\bar{u}$ is the mean value of u. Otherwise, this is rather a crude approximation in general, although not seriously misleading when e.g. $4N_e \bar{u}$ is small.

Can we use these results to test the neutral theory? As we shall see shortly, agreement between the observed and expected values of $\bar{H}$ will obtain only if we choose values of N_e grossly at variance with present-day numbers. Is this a mortal blow to the neutral theory? Not at all! We saw that the frequency of heterozygotes approaches its ultimate value very slowly. Suppose that for a long period in the past of a species under study, effective population size was small. Then the frequency of heterozygotes would attain a value appropriate to that small size. Although the population size may have greatly increased since then, the frequency of heterozygotes need not have increased very much in the time available. We can then obtain good agreement between the observed and expected values by postulating small effective population sizes in the past (Maynard Smith 1970). Note, however, that in view of the effects of migration, this could be the effective size of the whole species.

We can get a very rough idea of the appropriate sizes as follows (Kimura and Ohta 1971). From the rate of protein evolution and using a rough estimate of the proportion of amino-acid substitutions detectable

electrophoretically, we find the mutation rate per year to electrophoretically detectable mutants to be about 10^{-7} on average; the
corresponding value $\bar{v}$ for mutation rate per generation for any given
organism is then easily obtained. Our formula for H is not really suitable
for variants detected electrophoretically, since two chemically distinct
variants with the same electrical charge would (normally) perform
identically under electrophoresis; individuals heterozygous for two such
variants would be recorded as homozygotes. A more appropriate
formula has been obtained by Ohta and Kimura (1973) but for
$4N_e \times$ mutation rate less than about 0.2, the two formulae give much the
same answer. Let H^* be the observed mean frequency of heterozygotes
detected electrophoretically. For a rough answer (which is scarcely worth
refining in view of the rough estimates used) we put

$$H^* = \frac{4N_e\bar{v}}{4N_e\bar{v}+1}$$

and find $4N_e\bar{v}$ to be, for Man or mouse, in the region of 0.1. This gives
us, for effective population number in the course of evolution

Man   13 000   Mouse   500 000

Even if these are values for the whole species, they do not seem
unreasonable.

Thus our hero escapes again! However, further trials await the neutral
theory. To explain results in some species in terms of small population
sizes in the past is plausible; to invoke such "bottlenecks" in size
repeatedly in order to make the facts fit the theory would savour of
special pleading. Yet there seems little doubt that the frequency of
heterozygotes in large populations is substantially smaller than would be
predicted on any *strictly* neutralist model (Lewontin 1974, Selander 1976,
Soulé 1976). In fact, the average frequency of heterozygotes for variants
detected electrophoretically does not exceed 30%, although a much
larger value is expected for some species, save in the unlikely case that
the theoretical relationship between H^* and N_e for such variants is
hopelessly wrong. However, a minor modification of the neutral theory
(Ohta 1974, 1976) takes care of the difficulty. Suppose a fair proportion
of almost-neutral mutants are in fact very slightly disadvantageous. In
small or medium-sized populations, where drift has a much greater effect
than has minuscule selection, these mutants are effectively neutral and

the frequency of heterozygotes comes out as expected. In very large populations, however, the effect of drift is negligible even in relation to that of very small selection, and the selective disadvantage of our mutants makes itself felt. Thus the frequency of these mutants remains low, and heterozygosity is lower than expected on strict neutral theory. Note that on Ohta's theory, the rate of evolution would slow down whenever population size became very large, and this would help to explain why the rate varies.

Thus we have to invoke selection of some kind in order to explain allele frequencies. That this must be so, at least in the case of Man, was first shown by Haigh and Maynard Smith (1972). In outline their argument is as follows. Consider human haemoglobin and ignore variants known to be under selection. All other variants turn out to be individually very rare. If these variants are neutral, this rarity implies relatively recent origin; it may be shown that all the variant alleles, if neutral, must be copies of alleles arising by mutation during the past 500 or so generations (about 10 000 years). What then has happened to the descendants of mutant alleles which arose in the more distant past? Some of them should have got going, if they are neutral. This suggests that variant haemoglobin alleles are not neutral but mildly disadvantageous, frequencies being kept low by selection.

Alternatively, we can suppose that the alleles are neutral but that for some lengthy period in the past, the human population size was sufficiently small for all the ancient variants to have become extinguished. But if so, multiple allelic polymorphisms, if neutral, would have been eliminated (the effects of small human population size in the past are also discussed by Nei and Li 1975; they do not appear to contest this last point, although their scenario differs in some ways from that of Haigh and Maynard Smith). Multiple allele polymorphisms, then, such as are found for transferrins, would have to be attributed to natural selection or to a neutral mutation rate much higher than any yet calculated, on neutral theory, from the rate of evolution. The latter, although not impossible, seems very unlikely.

It appears then that we have to bring in natural selection somewhere. Ohta's theory does this but at the same time does minimal damage to the neutral theory. While we have to invoke selection in order to explain why many almost neutral variants have *not* spread in large populations, it is still possible to maintain that a substantial proportion of electrophoretically detected variants that *have* spread have done so purely by drift.

Testing the neutral theory from frequency data: summary

The reader will probably have concluded that most data currently available on genotype frequencies at loci controlling protein structure can be accommodated to the neutral theory, at least in the modified form proposed by Ohta, provided we are prepared to make the appropriate assumptions about unknown factors such as population numbers in the past, the extent of migration between sub-populations, and the frequency of very slightly disadvantageous mutants. However, this agreement between fact and theory is inconclusive; since the same facts can be explained on either the neutralist or the selectionist theory, the tests we have described enable no decision to be made between rival theories. Similar difficulties apply to all tests of the neutral theory proposed; none of them provides a critical criterion by which one of the theories can be rejected. Our ignorance of many relevant facts is such that any apparent objection to the neutral theory can be got round by making appropriate assumptions about unknowns. In the author's view, we cannot escape the depressing conclusion that ten years' arduous work, both on collecting data and on devising tests (often of great ingenuity) has led to no decisive answer. Many of these tests are reviewed critically by Ewens and Feldman (1976) and Ewens (1977). In a particularly disarming manner, these authors include tests devised by themselves among those that they criticize. Ewens concludes:

experimental techniques are still in a state of development, theoretical conclusions are incomplete and several years will pass until both have reached a form where rather reliable tests of the neutral theory, based on gene frequencies and population genetics theory, will have been arrived at.

Experimental approach

The author must confess to a sense of relief when turning from survey data to experiments. The relevant experiments have been of two kinds. In the first, several loci are studied simultaneously. Individuals are subjected to a number of environments, in the hope that if selection is operating genotype frequencies will come to differ from one environment to another (Powell 1971). The big advantage of this type of experiment is that even a comprehensive failure to detect selection will probably be reported; while success in finding selection at one or two loci should usually guarantee that the complete results will appear in the literature. This is important; we do not want our conclusions to be biased by a natural reluctance on the part of some workers to publish "failures" (there is nothing sinister about this reluctance, possibly the main factor is a feeling that negative results, which could always be due to the experiment having been carried out on too small a scale, are boring). The disadvantage of these experiments, as done so far, is that they do not distinguish between selection at the loci under study and selection at closely linked loci. We return to this point later.

In the second type of experiment, one locus is studied at a time. In the most interesting of these experiments, one starts with an outline study of biochemical properties of variants of a particular enzyme. One might find that variant A has optimum activity under one set of conditions encountered in nature by the species under study, whereas variant B has optimum activity under a different set of natural conditions. To demonstrate selection, we have to show that these differences in activity matter to the species concerned. Now in some cases we may plausibly assume as a working hypothesis that low activity is disadvantageous, so that under a given set of environmental conditions, the allele giving rise to enzyme with high activity is favoured. Random samples of individuals are exposed to these conditions for several generations in the laboratory, and we see if allele frequencies change in the predicted manner. If this happens, we can be confident that selection is operating at the locus concerned, with no worries about effects at linked loci. On the other hand, a failure to detect selection at a single locus may easily not be published.

The most promising approach would seem to lie in some combination of the two types of experiment. This has not yet been done; indeed, rather few studies of either type have been carried out. We shall describe a few here; for most of the remainder, see Lewontin (1974).

Experiments involving several loci

Pioneer studies along these lines were first made by Powell (1971). We shall discuss in detail an experiment by Minawa and Birley (1978), for which the results were particularly clear-cut. An eight-year-old laboratory cage population ("Texas") of *Drosophila melanogaster*, originating from 30 inseminated females collected from a natural population at Austin, Texas, was found to be polymorphic for eight out of 24 enzyme loci studied, variants being detected by starch gel electrophoresis. Two of these loci were difficult to study for technical reasons. Further investigation was therefore carried out on the remaining six polymorphic loci and on a locus polymorphic for the recessive mutant pink (eye).

Three environments, differing in respect of temperature regime and/or type of food supplied were used; two cages were set up for every environment, every cage being started with 2000 adults taken from the Texas population. Samples were taken from all six cages at 27 weeks after the start of the experiment and again 54 weeks later; genotype frequencies at the seven loci in these samples were determined on each occasion. Frequencies might vary with sample, either because of natural selection or because of random effects. The latter must contribute to differences between samples from *different* cages of the *same* environmental type. This forms the basis of a test of significance. If differences between samples from cages of *different* environmental type are significantly greater than differences between samples from different cages of the *same* environmental type, selection has been demonstrated.

Results were particularly striking. By the second occasion of sampling, five loci showed significant differences in genotype frequency between environments. These five loci, then, have responded differently to different environments. This rapid substantial divergence between environments implies intense selection.

The remaining two loci showed a significant difference in genotype frequency between the two occasions of sampling, in some environments; here again, the test of significance was designed in such a way that the detected differences between occasions must reflect the action of natural selection. Presumably changes at these two loci were roughly similar in all environments (although differing somewhat in magnitude) so that differences between environments proved non-significant.

Thus at all seven polymorphic loci (six of them enzyme loci), intense natural selection was operating. How typical are these results? In some of the experiments where the number of polymorphic loci studied was very small (e.g. one locus in the well-known experiment of Yamazaki 1971), no selection was detected. When, however, several enzyme loci were studied and results reported in detail (Sing, Brewer and Thirtle 1973, McDonald and Ayala 1974—all in *Drosophila*), selection at a fair proportion of loci was found. If selection at enzyme loci were unusual, it seems most unlikely that it would be so readily detected. More tentatively, one might suggest that, given a sufficiently wide range of environments, most loci would give evidence of selection. At any rate, the experiment would be well worth doing. It would, of course, be much the best if the environments chosen were reasonably representative of the environments encountered by the species in nature.

As already stated, these experiments do not show whether the selection was at the locus under study or at some linked locus. It could be that the alleles at a locus under study are (by chance) not associated at random with alleles at some linked locus or loci under selection. We discuss in the next chapter how this might happen. However, one point is clear. If the results given are indeed relevant to the situation in nature (a point which needs further investigation), changes in allele frequency at many enzyme loci arise through selection, whether at the locus concerned or a linked locus, drift playing a very minor role in these changes.

If a given locus responds in the same way to a given environment in experiments involving different populations, we may be confident that it is the locus itself which is responding to selection, since non-random associations with other loci brought about by chance should vary with population. If, however, response does vary with population, this could be due to selection at a linked locus *or* at the locus under study, the response at the latter being conditioned by interaction with selective effects at other loci, allele frequencies at such loci varying with population.

Predicting selective response from biochemical studies

One way, obvious in principle but not always easy in practice, to decide whether variants at a locus differ in fitness is to study the properties of the variants and attempt to predict which variant would convey the greatest selective advantage in a given environment. If the response in an actual experiment turns out as expected, we may be confident that selection is occurring at the locus concerned and not just at a linked locus or loci.

In the case of protein polymorphism, most of the work along these lines has been carried out on loci responsible for the activity of the enzyme *alcohol dehydrogenase* in *Drosophila melanogaster*. The pioneer work is due to Gibson (1970). Consider the locus coding for the enzyme structure. Two alleles $\underline{Adh^F}$ and $\underline{Adh^S}$ ($\underline{F}$ and $\underline{S}$ for short) are recognizable in most populations giving rise to structurally different molecules ADH-F, ADH-S detectable electrophoretically. Gibson found that the enzyme activity of $\underline{FF}$ larvae was about twice that of $\underline{SS}$ larvae, the heterozygote having intermediate activity (the greater activity being due to greater catalytic efficiency—Day, Hillier and Clarke 1974). He predicted that if the medium on which individuals were raised were supplemented with ethanol, the $\underline{F}$ allele would be at a selective advantage, the alcohol being more readily detoxicated (being converted to acetaldehyde) and metabolized. Gibson showed that $\underline{F}$ was indeed at an advantage in these circumstances, its frequency rapidly rising over 18 generations, while remaining constant over this time when no ethanol was added to the medium.

However, ADH-F was less stable on exposure to heat than ADH-S (the heterozygote was again intermediate). Thus at high temperature, $\underline{S}$ should be at an advantage over $\underline{F}$. Gibson suggested that in nature, the ethanol content of the habitat (rotting fruit) would fluctuate with strain of yeast. The combination of fluctuating temperature and fluctuating ethanol would give rise to a situation where sometimes one genotype, sometimes another, would be at an advantage.

A very large literature has grown up on this subject. For a recent account, see van Delden, Boerema and Kamping (1978). Although the situation is considerably more complicated than at first thought, it seems clear that the $\underline{F}$ and $\underline{S}$ alleles do react to selection by ethanol (and other alcohols) and to high temperature in the manner expected, at least in laboratory populations. The situation in nature is more difficult to assess. Briscoe, Robertson and Malpica (1975) found $\underline{F}$ frequencies in wine cellar populations to be much higher than in other Spanish populations, as expected. However, McKenzie and McKechnie (1978) have shown that alcohol tolerance in a natural population may result from selection at loci distinct from the locus determining enzyme structure, frequencies of $\underline{F}$ and $\underline{S}$ being unaffected by the presence of alcohol. It remains to be seen which method of producing alcohol tolerance is the most common in nature. At any rate, we have indicated a helpful line of approach. For similar studies on amylase see Scharloo, Dijken, Hoorn, de Jong and Thörig (1977).

Tentative conclusions

It is clear that the experimental results we have described present a serious challenge to the neutral theory *as usually stated*. The changes in allele frequency observed in such experiments are often much too rapid to be attributable to drift. As we have emphasized, it does not necessarily follow that the selection detected is acting on the loci under study, rather than on linked loci, although it should be possible ultimately to decide this point along the lines indicated for alcohol dehydrogenase. Further study will also be required to decide whether the experimental results are relevant to the situation in nature.

In the view of the present author, the case against the neutral theory, while not overwhelming, has become very formidable. This view is not, of course, shared by all. The reader may feel frustrated that we have not been able to reach a final decision after such a lengthy discussion. But as Swift has written:

> Wisdom is a hen, whose cackling we must value and consider, because it is attended with an egg. But then lastly, 'tis a nut, which, unless you choose with judgement, may cost you a tooth, and pay you with nothing but a worm (*A Tale of a Tub*).

Summary

Most mutants are harmful. Of those that are not, a significant fraction are thought by some population geneticists to be neutral, their fate depending on chance only. In the view of others, nearly all non-harmful mutants convey a selective advantage. If, as some think, such advantages are typically small, both selection and drift must be invoked in order to explain changes in allele frequencies at most loci. A third view is that the advantages are typically large, so that the effects of natural selection will usually overwhelm those of drift (save at extreme allele frequencies).

In the case of variants affecting morphology, even in a minor way, the dominating role of natural selection seems well established. Controversy in recent years has centred on the possible role of drift in the spread of variants at loci coding for enzyme structure, but with no known effects on morphology.

For a given protein, the rate of evolution has been roughly constant over different lines of descent. This finding, unexpected on the view that the rate depends mainly on natural selection, is easily explained on the assumption that most successful variants have spread by drift only. However, this explanation is sound only if the rate of mutation per year to neutral alleles at a given locus is the same in different lines of descent. Whether this is so is unknown.

It was hoped at one time that a decision on whether the frequencies of enzyme variants result from natural selection could be reached from a study of genotype frequencies in present-day natural populations. So far, this approach has not led to a decisive answer.

In laboratory experiments, frequencies of enzyme variants often change in a manner explicable only on the assumption of natural selection. It is uncertain whether the selection was occurring at the loci under study or at linked loci. Changes of frequency in such cases were determined almost entirely by selection, the effect of drift being very feeble in comparison. The relevance of these findings to events in the natural habitat remains to be determined.

In the author's opinion, the experimental approach offers the best hope for a final resolution of the controversy.

CHAPTER SEVEN

HITCH-HIKING EFFECT OF A FAVOURABLE GENE

Faster they rowed and faster,
But none so fast as stroke.

(author uncertain)

The hitch-hiking effect defined

In the last chapter we noted that a neutral allele could change in frequency as a result of linkage to a locus under selection. We now consider this point in a little more detail. Our aim will be to identify situations in which the change in frequency of the neutral allele is large. Fortunately, results accord very much with intuition; a brief discussion should therefore suffice.

Suppose A, a are neutral alleles at the same locus, both alleles being present in the population under study. At a linked locus, let there be alleles B, b; genotypes BB, Bb and bb are supposed to differ from one another in fitness.

If a is as common in chromosomes carrying B as in chromosomes carrying b, then whatever happens to B or b will have no effect on the frequency of A (or, of course, of a).

Things are quite different if the frequency of a among the B chromosomes differs from its frequency among the b chromosomes. Suppose, for example, a is over-represented among the B chromosomes, under-represented among the b chromosomes. We say, in such a case, that a is "associated" with B. Then as long as this association persists, changes in frequency of B will be accompanied by changes in frequency of a. Suppose selective advantages are such that B is rising in frequency (e.g. if BB is fitter than bb, with Bb intermediate in fitness). The rise in frequency of B will be accompanied by a rise in frequency of a; the neutral allele a has "hitched a lift" from the linked advantageous allele B.

This phenomenon is known (Maynard Smith and Haigh 1974) as the "hitch-hiking effect of a favourable gene".

Effect of recombination

The association we have described should gradually disappear, over successive generations, as a result of recombination. In smallish populations, the extent of association is affected by drift, but we shall ignore this complication here. We conclude that, *given enough time*, the association will disappear and hitch-hiking will stop. If fitnesses are such as to lead eventually to fixation of $\underline{B}$, there may not be enough time for the association to disappear completely before fixation occurs; this will be particularly the case when linkage is very tight. In such a case, hitch-hiking stops only when fixation occurs. In general, the looser the linkage, the earlier does hitch-hiking stop, other things being equal.

The set-up most favourable for hitch-hiking

How might the association arise? If $\underline{A}, \underline{a}$ have been neutral for a long time in the past (rather than having become neutral fairly recently owing to a change in environment) the association must be due to chance. We shall throughout consider only cases of chance association.

Suppose that the population starts off all $\underline{bb}$, but fitnesses are such as to favour a rise in frequency of $\underline{B}$. Now although the mutation $\underline{b} \rightarrow \underline{B}$ may happen several times, we know (chapter 4) that most of the resulting $\underline{B}$s will be lost by drift. It is, then, perfectly possible that only *one* of the $\underline{B}$s actually gets going and gives rise to *all* of the $\underline{B}$ alleles present in the population at a later date. Suppose that this successful $\underline{B}$ first appeared in a chromosome carrying $\underline{a}$. Then at the start, all $\underline{B}$ chromosomes are also $\underline{a}$ chromosomes and we are all set up for a successful hitch-hike. Of course if $\underline{a}$ were already very common, the effect of hitch-hiking would not be very dramatic; we are interested in the case where $\underline{a}$ is initially fairly low or intermediate in frequency and thus could travel far. It is clear, then, that if several $\underline{B}$s got going, at very roughly the same time, as might happen in a very large population, possibilities for hitch-hiking would be very limited, since some of these $\underline{B}$s would first appear in $\underline{a}$ chromosomes and some in $\underline{A}$ chromosomes, and association would not be established. We see, then, that hitch-hiking is most favoured when the population size is such that only one $\underline{B}$ mutant surmounts the possibility of chance loss.

What other conditions would lead to a large change in frequency of allele $\underline{a}$? If selection is such that $\underline{B}$ increases to a fairly high frequency fairly rapidly, opportunities for recombination during the rise of $\underline{B}$ will be limited. Thus the association between $\underline{a}$ and $\underline{B}$ will persist to some extent as $\underline{B}$ rises, and $\underline{a}$ will rise with $\underline{B}$ provided linkage is not so loose that the association is dissolved before $\underline{B}$ has risen much. We see then that hitch-hiking is most favoured when selection is intense and linkage tight. Consider, for example, selection on the haploid model. Write, as usual, α for the selective advantage of $\underline{B}$ over $\underline{b}$ and c for the recombination fraction. This case has been considered in detail by Maynard Smith and Haigh (1974). An exact theory (taking account of the effect of drift on the frequency of $\underline{B}$ when rare) seems very difficult to obtain but, as they note, their treatment is quite good enough to establish the main points at issue. Their results show that the effect of hitch-hiking on the frequency of $\underline{a}$ is negligible unless c is less than α. The effect of hitch-hiking, then, is very local. If, for example, the selective advantage is 1%, loci at a (recombinational) distance exceeding 1% will be virtually unaffected by hitch-hiking.

If, however, the recombination fraction is lower than the selective advantage, a very dramatic change at the $\underline{A}, \underline{a}$ locus may occur; in extreme cases near-fixation of a neutral allele ($\underline{a}$ in our example) will result. It is helpful to consider the details of the process. For the moment, we stay with the haploid model and continue to suppose that $\underline{B}$ first arose in a chromosome carrying $\underline{a}$. When $\underline{B}$ has frequency p, the increase of frequency of $\underline{B}$ in passing to the next generation will be about $\alpha p(1-p)$, if drift can be ignored (see chapter 4). On plotting this increase in frequency against p we find that initially (p small) $\underline{B}$ increases slowly; at a later stage (p intermediate) $\underline{B}$ increases much more rapidly (the increase being most rapid when $p = \frac{1}{2}$) only to slow down again as p becomes closer to unity. Owing to drift, changes will be rather quicker when p is near 0 or 1 than our formula implies, but the general pattern of change ("slow-quick-slow") is unaffected (Maruyama 1972, 1977).

What is happening to allele $\underline{a}$ during this time? Since, from the very start, there will be some $\underline{a}$ alleles which are not in a $\underline{B}$ chromosome, the increase in frequency of $\underline{a}$ will always be less than the increase in frequency of $\underline{B}$. This, however, will not prevent fixation of $\underline{a}$ in cases where recombination does not occur. It is, indeed, obvious that in the absence of recombination fixation of $\underline{B}$ must lead to fixation of $\underline{a}$; it makes no difference that $\underline{a}$ is changing more slowly than $\underline{B}$, since $\underline{a}$ starts off at a (usually much) higher frequency than $\underline{B}$ and so has less far to go to reach fixation. With recombination occurring, however, things are quite different; the increase in $\underline{a}$, in any generation, lags farther and farther behind the increase in $\underline{B}$. Thus in the earlier stages of the process, two opposing factors are acting on allele $\underline{a}$. On the other hand, $\underline{B}$ is increasing at an accelerating rate, so that the "pull" on $\underline{a}$ would, in the absence of recombination, steadily increase (until the frequency of $\underline{B}$ reaches $\frac{1}{2}$). On the other hand, the increase in $\underline{a}$ is falling ever behind the increase in $\underline{B}$. Thus for $\underline{a}$ to undergo a large increase in frequency in the long run, not too much recombination should occur between the time when $\underline{B}$ first appears and the time when $\underline{B}$ is beginning its period of rapid increase. As we noted earlier, a necessary condition for this is that c should be rather less than α. We can now see that two other considerations will enter: population size and dominance. If the population is very large, the initial frequency of $\underline{B}$ is very small. A long time, then, elapses before $\underline{B}$ reaches the point of rapid increase, giving many opportunities for recombination while $\underline{B}$ is still rare. With very large populations, then, c would have to be considerably less than α for much hitch-hiking to occur. As a rough guide, we may take the rather approximate formula given by Maynard Smith and Haigh for the case when c is less than about $\alpha/100$

$$\frac{\text{final frequency of } \underline{A}}{\text{initial frequency of } \underline{A}} = \frac{c}{\alpha}\log_e M$$

where M is the number of alleles at the $\underline{A}, \underline{a}$ locus (twice the population size in our case). We thus confirm, from a different argument, our earlier conclusion that hitch-hiking is of little importance in very large populations.

Finally we leave the haploid model and consider the effect of dominance. Over the initial period, when $\underline{B}$ is fairly rare, $\underline{BB}$ homozygotes are uncommon and the advance of $\underline{B}$ depends mainly on the fitness of the heterozygote $\underline{Bb}$. Partial or complete dominance of $\underline{B}$ over $\underline{b}$ in fitness should not seriously change our conclusions obtained from the haploid model. On the other hand, if the advantageous allele is recessive in fitness, its advance in the earliest stages, depending as it does on the presence of the very rare homozygote, will be exceedingly slow, giving many opportunities for association to be gradually dissolved.

The most favourable conditions for hitch-hiking then are (1) population size not too large, (2) recombination fraction very much less than the selective advantage of the heterozygote at the linked locus under selection. If the $\underline{a}$ allele is initially rare, its frequency will change most under hitch-hiking; but of course, if $\underline{a}$ is rare, it is very unlikely that $\underline{B}$ will first appear in a chromosome containing $\underline{a}$. In practice, then, a rise from a moderate to a very high frequency is the most that can be expected.

Some other situations leading to hitch-hiking

So far we have assumed that selection will lead to eventual fixation of the advantageous allele $\underline{B}$. Selection may, however, be such that $\underline{B}$ eventually settles down at some intermediate value. An example of this kind is discussed by Thomson (1977). Results are very much in line with those given above; if the final value of $\underline{B}$ is not too small, hitch-hiking will be substantial under conditions analogous to those we have given, although the overall effect on the frequency of $\underline{a}$ is, as expected, rather less than in the case where $\underline{B}$ goes to fixation. Thomson discusses many interesting aspects of the hitch-hiking phenomenon. Perhaps the most important of her findings concerns the effect of hitch-hiking on *two* loci, at neither of which are there differences in fitness, both closely linked to a locus under selection (this locus need not necessarily lie between the two "neutral" loci). She shows that non-random association of alleles at the neutral loci may be generated by hitch-hiking and persist for quite a long time. This is important; non-random association can be generated by selection (not involving hitch-hiking) and it might be supposed that if such association were found for alleles at two loci $\underline{A},\underline{a}$ and $\underline{C},\underline{c}$ this would provide evidence of selection at these two loci themselves (provided the population was large and random mating). Clearly, this is not necessarily the case.

A quite different model of hitch-hiking is discussed by Ohta and Kimura (1975). Instead of supposing that the advantageous mutant $\underline{B}$ arises in a chromosome carrying $\underline{a}$, they consider the case where neutral mutant $\underline{a}$ arises in a chromosome carrying advantageous mutant $\underline{B}$ en route to fixation. The analogous case when $\underline{B}$ settles down at an intermediate frequency is discussed by Thomson (1977). As expected, the effects of hitch-hiking on the frequency of neutral allele $\underline{a}$ are not very significant in these cases.

Summary

The frequency of a neutral allele $\underline{a}$ can change markedly by natural selection at a closely linked locus. The situation most favourable for this "hitch-hiking" occurs when an advantageous allele $\underline{B}$ (destined to be ultimately fixed) arises in a chromosome carrying a and is (by chance) the sole progenitor of all Bs present in the population thereafter. The change in frequency of a is most marked when the population is not too large and the recombination fraction is considerably less than the selective advantage of the Bb heterozygote. Unless the recombination fraction is less than the selective advantage, little change in frequency of a will obtain. Changes, although still striking, will be less in cases where B settles down at a roughly intermediate value than in cases where B is ultimately fixed. Non-random association of alleles at two neutral loci can be generated and persist for a long time if both loci are closely linked to a locus under selection.

CHAPTER EIGHT

CHANGES OF ALLELE FREQUENCY UNDER NATURAL SELECTION IN LARGE RANDOM MATING POPULATIONS

I'll tell you who Time ambles withal, who Time trots withal, who Time gallops withal, and who he stands still withal.

William Shakespeare, *As You Like It*

Conditions under which the problem can be simplified

In earlier chapters we have shown that changes of allele frequency under drift are very slow; rather little change will take place if the number of generations involved is substantially less than the effective population size. In contrast, intense selection will lead to very rapid changes over a small number of generations. When the intensity of selection is more modest, however, the outcome is not easily seen; it seems impossible to obtain an idea of the number of generations required to achieve a given change in allele frequency, unless the actual detailed calculations are carried out.

Now, in general, problems in which we are required to find the time for a given change are very difficult. Suppose allele frequencies are changing under drift as well as natural selection. If we imagine a set of populations of identical size, in which the set-up *vis-à-vis* selection is identical, and in all of which initial allele frequencies are the same, the number of generations required to reach a given allele frequency will still vary with population, owing to drift. At the very least, we should want to know the *mean* number of generations to achieve the given frequency. This can be found, but the analysis is quite difficult; we shall, therefore, when dealing with this general situation, confine ourselves to quoting and commenting on some leading results.

However, we know that in populations which are moderate or large in size, the effect of drift is very feeble when allele frequencies are intermediate. We surmise that unless the intensity of selection is very low, the effect of natural selection will overwhelm that of drift in such cases; we should not go far wrong by treating the problem "deterministically", that is, by ignoring the effects of drift completely. This greatly simplifies the analysis and gives results valid under a wide range of circumstances. We shall discuss these circumstances in more detail later, but in order to illustrate the limitations of the deterministic approach we consider briefly the case of genic selection; as before, let α be the selective advantage of $\underline{A}$ over $\underline{a}$ and let N be the population size. In chapter 4, we showed that the effect of drift on probability of fixation fades out once the frequency of the advantageous allele has reached about $2/(N\alpha)$. It would seem reasonable to guess that from this frequency onwards drift will not have a serious effect on the rate of change of allele frequency until such time as the frequency of the favoured allele is fairly close to unity. This surmise is indeed correct, as was shown by Ewens (1967, 1969). Our value $2/(N\alpha)$ is a little stringent; the effect of drift becomes rather trivial once the frequency of the advantageous allele reaches $1.5/(N\alpha)$. Ewens shows that, for genic selection, drift becomes important again when the frequency reaches about $1-1.5/(N\alpha)$. To put the matter more strictly: within the range $1.5/(N\alpha)$ to $1-1.5/(N\alpha)$, the *mean* number of generations required to achieve a given change is virtually the same as the corresponding number of generations calculated deterministically. In principle, we still have the difficulty that drift will cause deviations from this mean. In practice this is not a serious problem, for the following reason. If the deterministic approach is to be of much use, we should want it to apply over a substantial part of the range of allele frequencies, say 0.005 to 0.995. For $1.5/(N\alpha)$ to equal 0.005 (or less), we must have $N\alpha = 300$ (or more). With this large value of $N\alpha$, the mean number of generations required to go from 0.005 to 0.995 is substantially less than the population size, so that over this range deviations from the mean due to drift will be rather small. Thus the deterministic approach will be reliable over this range, provided the population is large enough for $N\alpha$ to equal 300 or more. For $\alpha = 0.005$, this means a population size of 60 000 or more.

On the other hand, if allele frequencies are very extreme, the population size must be really large if the deterministic approach is to give reliable results. If we want $1.5/(N\alpha)$ to be as low as $1/10 000$, we must have $N\alpha = 15 000$, that is, a population size of 3 million if α is as low as 0.005; the more extreme the frequencies, the larger would have to be the population size. We stress, however, that no matter how large the population size, it is *not* possible for the deterministic approach to give reliable results for the *complete* range $(1/(2N)$ to 1) of frequencies.

For the moment we shall suppose that the population size is sufficiently large for deterministic theory to apply over the range of allele frequencies we shall consider. Our treatment is essentially that of Haldane (1924, 1932, 1932a).

Genic selection

The easiest case to deal with is genic selection. As in chapter 4, write the (relative) viabilities of alleles A, a as:

	A	a
Viability	$1+\alpha$	1

Let the frequencies of alleles A, a in some generation be p, q and in the following generation be p^*, q^*. We have then, as in chapter 4

$$p^* = \frac{p(1+\alpha)}{1+\alpha p}$$

so that

$$q^* = 1 - p^* = \frac{q}{1+\alpha p}$$

whence

$$\frac{p^*}{q^*} = \frac{p}{q}(1+\alpha)$$

Thus the ratio of the two frequencies increases by a factor $(1+\alpha)$ every generation. Suppose initial frequencies are p_0, q_0 and frequencies in any generation t are p_t, q_t. Let $u_t = p_t/q_t$. Then

$$u_1 = (1+\alpha)u_0, u_2 = (1+\alpha)u_1 = (1+\alpha)^2 u_0$$

and generally

$$u_t = (1+\alpha)^t u_0$$

$$\log_e u_t = \log_e u_0 + t\log_e(1+\alpha)$$

$$t = \frac{\log_e u_t - \log_e u_0}{\log_e(1+\alpha)} \quad \text{generations}$$

Suppose, for example, $\alpha = 0.01$. How many generations are required for the frequency of $\underline{A}$ to change from $1/10\,000$ to $9999/10\,000$? We have

$$p_0 = 0.0001, \quad q_0 = 0.9999, \quad u_0 = 0.0001$$
$$p_t = 0.9999, \quad q_t = 0.0001, \quad u_t = 9999$$

whence

$$t = 1851 \text{ generations}$$

to the nearest whole number.

For $\alpha = 0.005$ and $\alpha = 0.05$ the corresponding values of t are 3693 generations and 378 generations respectively. Clearly these changes are occurring in a time shorter by several orders of magnitude than would be required for comparable changes in populations of the same size under drift.

The treatment just given is exact. In cases other than genic selection, however, an exact formula for the number of generations has not been obtained (save for very special values of selective advantage). If we want a formula, we shall have to resort to approximations. These often worry beginners, who feel justifiably uncertain as to the conditions under which such approximations are appropriate. It may help, therefore, if we illustrate here a standard approximate procedure which we shall use later, since the accuracy of the approximation is readily checked in the present case.

We saw in chapter 4 that if α is sufficiently small we have the close approximation

$$p^* - p = \alpha pq$$

Write $p^* - p = \delta p$; this is the change in allele frequency over time $\delta t = 1$ generation. Thus we may write

$$\frac{\delta p}{\delta t} = \alpha pq$$

provided α is small. By repeated application of this formula, we could calculate the allele frequency p_t in any generation t, once given the initial frequency p_0. Having done this, we could plot p_t against t. Suppose we passed a smooth curve through the set of points so obtained. If we had been able to establish the equation of this curve, without having to plot the points first, we could have calculated the value of p_t for any given t very readily.

Suppose "evolution is slow", that is, α is small. Then the allele frequency changes little over a few successive generations, so that δp, although in principle dependent on p, is much the same for a few successive generations; in the neighbourhood of any given t, the curve is virtually a straight line. Hence, near enough

$$\frac{dp}{dt} \text{ for the curve} = \frac{\delta p}{\delta t}$$

at any time t, provided α is small. Then

$$\frac{dp}{dt} = \alpha pq$$

and the time required to go from p_0 to p_t is

$$t = \int_{p_0}^{p_t} \frac{1}{\alpha pq} dp = \frac{1}{\alpha} \int_{p_0}^{p_t} \left\{\frac{1}{p} + \frac{1}{q}\right\} dp$$

$$= \frac{1}{\alpha} [\log_e p - \log_e q]_{p_0}^{p_t}$$

$$= \frac{1}{\alpha} \left[\log_e \frac{p}{q}\right]_{p_0}^{p_t}$$

$$= \frac{\log_e u_t - \log_e u_0}{\alpha} \quad \text{generations}$$

This is the same as the exact formula, except that we have α instead of $\log_e(1+\alpha)$. Now (appendix 2)

$$\log_e(1+\alpha) = \alpha - \tfrac{1}{2}\alpha^2 + \tfrac{1}{3}\alpha^3 - \tfrac{1}{4}\alpha^4 + \dots$$

so that when α is small, the two formulae are virtually identical. The approximate formula gives an underestimate of the value of t, but unless α is fairly large, the underestimate if fairly slight. Defining the percentage error as

$$\frac{t_{\text{exact}} - t_{\text{approximate}}}{t_{\text{exact}}} \times 100$$

we find

α	*Error* *(per cent)*
0.01	0.5
0.05	2.4
0.10	4.7

This approximate method, then, gives quite satisfactory answers in this case for α up to 0.05 or so.

Wright's Formula

Consider now a more general case:

	AA	Aa	aa
Frequency among (newly formed) zygotes	p^2	$2pq$	q^2
Viability	a	b	c

The average viability $\bar{w}$ is (by definition of average)

$$ap^2 + 2bpq + cq^2$$

One generation later, the frequency of $\underline{A}$ among newly formed zygotes will be

$$p^* = \frac{ap^2 + bpq}{\bar{w}} = \frac{p}{\bar{w}}(ap + bq)$$

Thus

$$ap + bq = \bar{w}p^*/p$$

and similarly, putting $q^* = 1 - p^*$ we find

$$bp + cq = \bar{w}q^*/q$$

From the usual rules for differentiation, bearing in mind that $q = 1 - p$, we find

$$\frac{1}{2}\frac{d\bar{w}}{dp} = (ap + bq) - (bp + cq)$$

$$= \bar{w}\left(\frac{p^*}{p} - \frac{q^*}{q}\right)$$

$$= \frac{\bar{w}}{pq}(p^*q - pq^*)$$

Write δp for $p^* - p$, the change in frequency of $\underline{A}$. On substituting $q = 1 - p$, $q^* = 1 - p^*$ we find

$$p^*q - pq^* = \delta p$$

and we have finally Wright's formula

$$\delta p = \frac{pq}{2\bar{w}}\frac{d\bar{w}}{dp}$$

(Wright 1937). The rather attractive proof we have given is taken, with minor modifications, from Edwards (1977).

With mild selection the viabilities a, b, c will be much the same in value and since, as explained in chapter 4, only relative viabilities matter in the present context, we may put a, b and c each close to unity. Thus $\bar{w}$ is approximately unity. Hence with mild selection we may use the approximation $\bar{w} = 1$. We can see the analogy here to our procedure with genic selection by noting that if we write viabilities, as in chapter 4, as

$$1 + s, 1 + sh, 1$$

instead of a, b, c then

$$\frac{d\bar{w}}{dp} = s \times \text{terms in } p \text{ and } h$$

$$\bar{w} = 1 + s \times \text{other terms in } p \text{ and } h$$

If we note (appendix 2) that for any x numerically less than unity

$$(1+x)^{-1} = 1 - x + x^2 - x^3 + \ldots$$

we see that replacing

$$\frac{1}{\bar{w}} \frac{d\bar{w}}{dp} \quad \text{by} \quad \frac{d\bar{w}}{dp}$$

is the same as ignoring terms in $s^2, s^3, s^4, \ldots$ which is of course justified with mild selection.

We have then the approximation, valid for mild selection

$$\delta p = \tfrac{1}{2} pq \frac{d\bar{w}}{dp}$$

and a little later we shall, as with genic selection, use the further approximation

$$\delta p = \frac{dp}{dt}$$

in order to calculate the number of generations required for the frequency of the advantageous allele to change from p_0 to p_t. However, we can obtain a fairly comprehensive view of the whole process of change in allele frequency from low value to high merely by considering the magnitude of δp for some special cases, which we now discuss.

Change in allele frequency per generation (mild selection)

With <u>AA</u> at an advantage to <u>aa</u> but with no dominance in fitness, we have

	A<u>A</u>	<u>A</u>a	aa
Viability	$1+2\alpha$	$1+\alpha$	1

Thus

$$a = 1+2\alpha, \quad b = 1+\alpha, \quad c = 1$$

$$\frac{1}{2}\frac{d\bar{w}}{dp} = p(a-b)+q(b-c) = \alpha p + \alpha q$$

$$= \alpha$$

Hence

$$\delta p = \alpha pq$$

to a good approximation. Of course, this is the same result as with genic selection, since α is small.

With <u>AA</u> at an advantage to <u>aa</u> and with complete dominance in fitness we shall, for purposes of comparison with the no dominance case, adopt the convention that the homozygote <u>AA</u> is as viable in the case of complete dominance as it was with no dominance; that is, we write for the case <u>A</u> dominant to <u>a</u>

	A<u>A</u>	<u>A</u>a	aa
Viability	$1+2\alpha$	$1+2\alpha$	1

Thus with our convention, the heterozygote <u>A</u>a "gains" from the presence of dominance, having viability $(1+2\alpha)$ rather than $(1+\alpha)$. We have then

$$a = b = 1+2\alpha, \quad c = 1, \quad \frac{1}{2}\frac{d\bar{w}}{dp} = 2\alpha q$$

$$\delta p = 2\alpha pq^2$$

when α is small.

When <u>aa</u> is at an advantage to <u>AA</u>, but with <u>a</u> still recessive in fitness to <u>A</u>, we shall continue to write p for the frequency of the *advantageous* allele.

Hence we must write

	AA	Aa	aa
Zygotic frequency	q^2	$2pq$	p^2
Viability	c	b	a

with

$$a = 1 + 2\alpha, \quad b = c = 1, \quad \frac{1}{2}\frac{d\bar{w}}{dp} = 2\alpha p$$

$$\delta p = 2\alpha p^2 q$$

when α is small.

We ask:

(1) For what values of p is evolution most rapid?
(2) What is the effect of dominance?
(3) How many generations to get from p_0 to p_t?

To answer (1) and (2) consider the special cases p small (that is, when the advantageous allele is still rare) and q small (that is $p = 1 - q$ is near 1, so that the advantageous allele is very common). When p is small, q is near 1, so that we may put $q = 1$ without risk of serious error, and similarly we put $p = 1$ for the case where p is close to 1. This gives us the approximate formulae presented in table 8.1. We shall refer to the three situations (i), (ii), (iii), as the dominant, intermediate and recessive cases respectively.

Table 8.1 Change in frequency of a mildly advantageous allele in a single generation (formulae slightly approximate).

	δp	δp when p is small	δp when p is near 1
(i) $\underline{AA}$ advantageous $\underline{A}$ dominant to $\underline{a}$	$2\alpha pq^2$	$2\alpha p$	$2\alpha q^2$
(ii) $\underline{AA}$ advantageous no dominance	αpq	αp	αq
(iii) $\underline{aa}$ advantageous $\underline{a}$ recessive to $\underline{A}$	$2\alpha p^2 q$	$2\alpha p^2$	$2\alpha q$

Initially, when the advantageous allele is still rare, p is small and therefore δp is small in all cases. Initially, then, p changes slowly.

At the end, when the advantageous allele has become common, q is small and therefore δp is again small in all cases. Thus at the end p changes slowly.

Consider the intermediate case. We have

$$\frac{d(\delta p)}{dp} = \frac{d}{dp}(\alpha p - \alpha p^2) = \alpha(1 - 2p)$$

$$\frac{d^2(\delta p)}{dp^2} = -2\alpha$$

When $p = \frac{1}{2}$,

$$\frac{d(\delta p)}{dp} = 0, \quad \frac{d^2(\delta p)}{dp^2} \text{ is negative}$$

so that δp is maximal when $p = \frac{1}{2}$. Similarly, for the dominant case δp is maximal when $p = \frac{1}{3}$, for the recessive case when $p = \frac{2}{3}$. Thus in all cases, change is initially slow, is most rapid at some intermediate frequency, and is again slow towards the end.

However, the details of the process are very much affected by dominance. Whereas both in the dominant and in the intermediate case δp in the early stages is proportional to p, so that change is "slow" but not "very slow", in the recessive case δp in the early stages is proportional to p^2; when p is small, p^2 is very small, so that δp is very small indeed. Hence an allele advantageous but recessive in fitness, even if it eventually spreads, will spend a very long time at very low frequencies before it becomes at all common. This is not surprising, since the advantage appears only in homozygotes, whereas in the early stages the allele is represented almost entirely in heterozygotes. Now the selective advantage may not persist throughout this long period, since the environment may change in such a way that the advantage is destroyed. Moreover, during this long period at low frequencies, the allele may well be lost by drift. Indeed, in a large population, size N, its probability of fixation is only about

$$1.128 \sqrt{\frac{2\alpha}{N}}$$

(Kimura 1957). Any calculations for the recessive case which ignore drift are therefore purely formal when we are considering these early lengthy stages, although for later stages the calculations give a fair representation of the process, provided the population is not too small and the allele frequency not too close to unity.

In principle, δp is very small in the dominant case towards the end of the process, when p is close to 1, but this is not very important. It makes no real difference to the evolution of the species whether the frequency of the advantageous allele is, say, 0.99 rather than 1. In any case, even if the frequency reaches unity, it will not stay there for long, owing to mutation to disadvantageous alleles. Moreover, at these late stages, our calculations are subject to the population being sufficiently large for us to ignore the effect of drift, as we explained earlier. The larger the population, the nearer will p approach 1 before drift begins to affect the outcome, but for no finite population can the effect of drift be ignored for *all* values of p close to unity.

We have stressed these points about drift, because it is sometimes assumed by beginners (incorrectly) that the effect of drift on change of frequency under selection can always be ignored provided the population is sufficiently large. Unfortunately, a more exact statement, although obtainable (Maruyama 1977) will probably strike the less mathematical reader as very difficult to grasp. Such a reader is advised to proceed to the next paragraph; for those who wish for a more formal statement, however, the point amounts to the following. For any value of p, *once stated*, it is possible to find an N sufficiently large for drift to be ignored, but the value of p must be stated *before* we calculate N. This is not at all the same as saying that a sufficiently large N can be found such that drift can be ignored for *all* p; such a (finite) N does not exist. In technical language, "convergence is not uniform". Readers who find this notion obscure will be relieved to discover that much mental anguish was undergone by nineteenth-century mathematicians before the idea and its significance in many problems was discovered. A technical but very enjoyable historical account is given in Lakatos (1976, Appendix 1).

We have set down the essential points. However, it is helpful to calculate the actual number of generations required for a given change in allele frequency. We shall now discuss how this is done.

Time required for a given change

We continue to use the deterministic approach. For the dominant case we have, to a good approximation

$$\frac{dp}{dt} = 2\alpha p q^2$$

provided α is small.

Now, since $q = 1 - p$, we have

$$\frac{1}{p} + \frac{1}{q} + \frac{1}{q^2} = \frac{1}{pq^2}$$

whence the number of generations to go from p_0 to p_t is

$$t = \int_{p_0}^{p_t} \frac{1}{2\alpha pq^2} \, dp = \frac{1}{2\alpha} \int_{p_0}^{p_t} \left\{ \frac{1}{p} + \frac{1}{q} + \frac{1}{q^2} \right\} dp$$

$$= \frac{1}{2\alpha} \left[\log_e p - \log_e q + \frac{1}{q} \right]_{p_0}^{p_t}$$

$$= \left(\log_e u_t - \log_e u_0 + \frac{1}{q_t} - \frac{1}{q_0} \right) \Big/ 2\alpha$$

$$= (\log_e u_t - \log_e u_0 + u_t - u_0)/2\alpha \text{ generations}$$

(where, as usual, $u_t = p_t/q_t$ and $u_0 = p_0/q_0$ so that $1/q_t = 1 + u_t$, $1/q_0 = 1 + u_0$).

Numerical values for the case $\alpha = 0.005$ are given in table 8.2.

Table 8.2 Number of generations required to bring about a given change in allele frequency when the advantageous allele is dominant, intermediate or recessive in fitness. Selective advantage $\alpha = 0.005$.

	Number of generations		
Change in allele frequency	Dominant	Intermediate	Recessive
0.0001 to 0.01	463	923	990 462
0.01 to 0.1	250	480	9 240
0.1 to 0.3	167	270	802
0.3 to 0.5	142	170	218
0.5 to 0.7	218	170	142
0.7 to 0.9	802	270	167
0.9 to 0.99	9 240	480	250
0.99 to 0.9999	990 462	923	463

The formula we have given is a little approximate but, from a comparison with a much more exact formula given by Haldane (1932), it appears that the error in our results is trivial in the case $\alpha = 0.005$. Indeed if as before we define the percentage error as

$$\frac{t_{\text{exact}} - t_{\text{approximate}}}{t_{\text{exact}}} \times 100$$

it appears that for α up to about 0.025, the percentage error will hardly exceed 200α. The error is greatest for values of p near 1. We can (following Haldane) improve the accuracy for such values of p, without seriously affecting the accuracy for other values of p, by replacing 2α in our formula by

$$k = \frac{2\alpha}{1+2\alpha}$$

With this modification, we can safely use our formula for all cases where α does not exceed 0.025; the percentage error will be (at most) about 100α. Of course, in practice, these errors are trivial in comparison with errors introduced by using an inaccurate value of α; it is indeed a formidable task to obtain reliable estimates of selective advantages operating in nature.

The recessive case may be dealt with in the same way as in the dominant case. The formula for t comes out the same as in the dominant case, provided we *redefine* u_t as q_t/p_t, u_0 as q_0/p_0 and replace 2α by -2α. Numerical results are given in table 8.2. The percentage error, for α not exceeding 0.025, will be at most about 100α; our formula should *not* be modified in the way we did in the dominant case.

When α exceeds 0.025 or so, it is best to use Haldane's formula (Haldane 1932)

$$t = \frac{u_t - u_0}{k} + \frac{\log_e\left(\dfrac{1+1/u_t}{1+1/u_0}\right)}{\log_e(1-k)} + \frac{1-k}{k}\log_e\left(\frac{1+u_t}{1+u_0}\right)$$

where
$$u_t = p_t/q_t, \quad u_0 = p_0/q_0, \quad k = \frac{2\alpha}{1+2\alpha} \quad \text{in the dominant case}$$
and
$$u_t = q_t/p_t, \quad u_0 = q_0/p_0, \quad k = -2\alpha \quad \text{in the recessive case}$$

(anyone wishing to read Haldane's paper should note that, in his convention, q and not p is the frequency of the advantageous recessive allele). The derivation of this formula is complicated and will not be given here.

For the case of no dominance, the approximate formula is the same as the approximate formula for genic selection. Numerical values, for $\alpha = 0.005$, are given in table 8.2. For an improved formula, replace α by $\alpha/(1+\alpha)$; once again this gives good results for α up to 0.025. Cases of incomplete dominance may be dealt with in an analogous manner, provided selective advantages are small. Intense selection presents more of a problem; straightforward formulae analogous to those for the dominant and recessive cases have not been found, even for the case of no dominance. A solution to the problem is given by Haldane and Jayakar (1963), but unfortunately the procedure is cumbersome. Now as Kimura (1968*a*) has pointed out in an interesting article, Haldane seems to have been particularly intrigued with this problem. Were there a simple solution, it seems likely that Haldane would have found it, so that we can be fairly certain that no simple solution exists. This being so, the best procedure in these cases is to find the exact formula for δp by substituting in Wright's formula and (by repeated use of the formula for δp) evaluate on a computer the frequency of the advantageous allele in successive generations.

Allowances for drift

We consider now the effect of drift. In particular, we shall be interested in the *mean* number of generations spent by the advantageous allele in a given range of frequencies, x to y, say. Obviously the answer will depend on the initial frequency; we shall take this to be $1/(2N)$ in all cases.

Now it makes a great difference whether we consider (i) *all* mutants of the same selective advantage and initial frequency or (ii) confine ourselves solely to those mutants, of same advantage and initial frequency, *destined ultimately to be fixed*. With initial frequency $1/(2N)$, most mutants are lost by drift before they reach frequencies which are at all large; the average time spent by mutants as a whole at high frequency will be quite small, since the many zero times are averaged in. On the other hand, mutants which are eventually fixed will obviously spend a fair time at high frequencies.

We confine ourselves here entirely to advantageous mutants which are ultimately fixed, this being the relevant case in the present context. Suppose we have mild genic selection. The mean number of generations spent by the advantageous allele in the range of frequencies x to y turns out to be

$$\int_x^y \frac{1 - e^{-4N\alpha p} - e^{-4N\alpha(1-p)} + e^{-4N\alpha}}{\alpha p(1-p)(1 - e^{-4N\alpha})} \, dp$$

(Maruyama 1972, 1977; Ewens 1973). In general, this will have to be evaluated numerically on a computer. The formula looks complicated but, for $N\alpha$ sufficiently large, a substantial simplification can be achieved. Firstly, the term $e^{-4N\alpha}$ may be ignored when $N\alpha$ exceeds 2. Secondly, the term $e^{-4N\alpha p}$ may be ignored when p exceeds $2/(N\alpha)$. Thirdly, the term

$e^{-4N\alpha(1-p)}$ may be ignored when p is less than $1-2/(N\alpha)$. We see then that provided p lies in the range

$$2/(N\alpha) \text{ to } 1-2/(N\alpha)$$

our formula for the mean number of generations reduces to the formula for the number of generations given by the deterministic approach. As mentioned earlier, the latter approach will, near enough, give the mean for the range $1.5/(N\alpha)$ to $1-1.5/(N\alpha)$, as was shown by Ewens (1967) using a slightly different approach.

Our main interest will be to calculate the mean number of generations spent in the complete range $1/(2N)$ to 1. To avoid awkward complications, we consider the range $1/(2N)$ to $1-1/(2N)$; this makes a difference of only 2 generations to the answer. Suppose now that $N\alpha$ exceeds 20. Over most of the range $1/(2N)$ to $2/(N\alpha)$, $(1-p)$ will then differ little from unity. Thus the "correction" to be subtracted from the deterministic result to allow for the effect of drift when p is small will be approximately

$$\int_{1/2N}^{2/N\alpha} \frac{e^{-4N\alpha p}}{\alpha p} dp$$

which may be shown to equal (near enough)

$$(-0.5772 - \log_e 2\alpha + 2\alpha)/\alpha$$

provided α is small. An identical correction must be subtracted to allow for the effect of drift when p is near 1.

We stress that these corrections are *not* trivial. Suppose, for example, $\alpha = 0.005$; the correction comes out equal to 807.6, so that the mean time over the whole range is about 1615 generations less than the value given by the deterministic result. Suppose the population is size 1 000 000. For $\alpha = 0.005$, the number of generations to go from frequency $1/(2N)$ to $1-1/(2N)$ under genic selection comes out to be 5803 generations using the approximate deterministic formula, whereas we see that the correct answer is about $5803 - 1615 = 4188$ generations.

The discrepancy becomes particularly marked in cases where the advantageous allele is completely dominant in fitness. Given $N\alpha$ fairly large, the correction for drift when p is small will be much the same as for genic selection, remembering however that if we adhere to our convention of writing the fitness of the heterozygote as $(1+2\alpha)$ in cases of complete dominance, we must write 2α for α and 4α for 2α in our formula for the correction. For values of p near unity, however, the deterministic approach gives results quite grotesquely wrong. Thus (Ewens 1967, 1969) when $2\alpha = 0.001$, $N = 10^6$, deterministic theory tells us that to go from $p = 0.999$ to $p = 0.99999$ takes 99 million generations, whereas the correct mean time is 4000 generations. In fact, the deterministic approach becomes unreliable once the frequency of the advantageous allele exceeds

$$1 - \frac{5}{\sqrt{4N\alpha}}$$

which is 0.89 in the case $2\alpha = 0.001$, $N = 10^6$. For further details see Ewens (1969). As we indicated earlier, similar very large discrepancies will be found in cases of advantageous mutants recessive in fitness when p is near zero. Almost all such mutants will be lost by chance, but the odd mutant that gets going will move through the early stages very much faster than deterministic theory indicates.

Industrial melanism

To paraphrase Macaulay, every schoolboy knows that dark moths are at a selective advantage over light moths in areas where industrial air pollution has blackened tree trunks and killed off lichens, whereas in unpolluted areas light are at a selective advantage over dark. In a famous series of investigations on the Peppered Moth *Biston betularia* carried out in the 1950s, Kettlewell observed directly the proportion of moths of either type taken from tree trunks by bird predators; moreover, he released marked moths of both types and later recaptured a random sample of his releases. Results from these two approaches were in agreement; the recapture data showed that dark were (about) twice as likely to survive as light in a heavily polluted area, whereas light were (about) twice as likely to survive as dark in an unpolluted area. This brief summary cannot do justice to what is rightly regarded as a model of careful investigation of natural selection in action; the reader in search of enlightenment is strongly recommended to read Kettlewell's book (1973).

These findings are sometimes regarded by beginners as a complete explanation not only of the rise of dark forms of so many species of moths since the middle of the nineteenth century but also of their present-day distribution. In virtually all of these species, dark is dominant to light. The spread of the allele conferring melanism would then follow the theoretical scheme set out earlier in this chapter for advantageous dominants. Those who take this view may be surprised to discover that they are *plus royaliste que le roi*; Kettlewell has repeatedly pointed out that such a simple explanation does not cover all the facts. While agreeing, of course, that cryptic colouration is a major factor in the rise and maintenance of industrial melanism, we shall briefly indicate some problems well known to those working in this area. For a particularly helpful account, see Bishop and Cook (1975).

One difficulty in the case of the peppered moth, obvious from the start, is the existence of a third form *insularia*, variable but intermediate in form to dark (*carbonaria*) and light (*typica*); *insularia* is dominant to *typica* but recessive to *carbonaria*. Although, at least in Britain, *insularia* is rather infrequent or even absent in many areas, this is not always the case, and in some areas it is the most common of the three forms (Kettlewell 1973, Appendix C). No generally agreed explanation for the frequency of *insularia* has been achieved.

A more serious difficulty is the presence of the "wrong" form, at least (but not always) at low frequency. As Kettlewell points out, it is very unusual for the *carbonaria* phenotype to exceed 98 % in frequency, even in a heavily polluted area. The first quantitative treatment of the spread of melanism was made by Haldane (1924). He assumed, plausibly, that the frequency of the *carbonaria* phenotype at the time when the first specimen was found in Manchester in 1848 was about 1 %, and showed that for the frequency to rise to 99 % in about 50 years the fitness of the *carbonaria* phenotype to that of the *typica* phenotype would be about 1.5:1. To put the matter another way, a selective advantage of this magnitude would be sufficient to raise the frequency to 99 % in about 50 years (this figure is not altered much if the initial frequency, assumed at 1 %, was in fact as low as 0.1 %). Since Kettlewell's experiments show an advantage of 2:1 rather than 1.5:1, the frequency of *typica*, prior to the adoption of clean air policies, should have fallen to minuscule proportions.

It is, of course, possible that these *typica* are mainly migrants from non-polluted areas; migration from polluted areas could similarly explain cases where a minority of melanics are found in an unpolluted area. But this is not the whole story; a large *majority* of melanics are sometimes found in an area where experiments show *typica* at an advantage *in respect to bird predation* (Lees and Creed 1975). It seems then that bird predation is not the complete explanation for the frequencies of melanics; these must have advantages other than those related to cryptic colouration. It seems that only if the disadvantage of the melanics in respect to bird predation is sufficiently large to outweigh these other advantages will melanics fall to a low frequency. This is presumably the explanation for the unexpectedly high frequency of *carbonaria* found in part of North Wales (Bishop 1972).

Equally surprising are differences between species in respect of frequency of melanics. Bishop and Cook (1975) studied the frequency of melanics along a 30-mile corridor extending from industrial North-Western England into rural North Wales in three species of moth: *Biston betularia, Phigalia pilosaria* (pale brindled beauty) and *Gonodontis bidentata* (scalloped hazel). In all three species, melanics were much more common in industrial than in rural areas. However, in all areas the frequency of melanics in *Biston* exceeded that in *Gonodontis*, often by a large margin; *Phigalia* frequencies were intermediate between those of the other two species. *Gonodontis* migrates very much less than *Biston*, so that the lower frequency in industrial areas cannot be ascribed to greater migration from non-industrial areas. If visual advantages and disadvantages are about the same in all three species, we must postulate rather large differences between species in non-visual effects. For further discussion of these and other problems of industrial melanism, the reader is referred to the publications we have cited.

Summary

A newly arisen advantageous allele which survives chance loss increases initially rather slowly, particularly when the allele is recessive in fitness. As the allele becomes more common, the rate of change increases markedly, only to fall again as the allele becomes more and more abundant. When the allele is dominant in fitness, change is very slow when the allele is very common. Drift will substantially increase the rate of change at extreme frequencies, but in a large population has little effect at intermediate frequencies unless the selective advantage is very small.

When selective advantage is really large, allele frequencies change very rapidly, as in the case of industrial melanism. Although the rise and present-day frequency of melanics is, in the main, a reflection of their concealment from bird predators under conditions resulting from industrial air pollution, other forms of selection seem to be acting on the melanic phenotype.

CHAPTER NINE

EVOLUTION OF DOMINANCE

The pronounced tendency of the mutant gene to be recessive, to the gene of wild type from which it arises, calls for explanation.

Sir Ronald Fisher, *The Genetical Theory of Natural Selection* (1930)

Modifiers of dominance

Hitherto, we have treated dominance as a fixed property of the alleles concerned, both in cases where we were discussing dominance in fitness and in cases where we discussed dominance at the gross phenotypic level, as in the case of industrial melanism. However, there is much experimental evidence to indicate that the degree of dominance of one allele over another at a given locus (the "main" locus) can be affected by the presence or absence of specific alleles ("modifiers of dominance") at other loci. Ford (1940) showed that, in the currant moth, yellow wing ground colour could be made either near-dominant or near-recessive to white ground colour by selection of the appropriate genetic background. Fisher and Holt (1944) showed that, in the mouse, the mutant Danforth's short tail, which normally shows marked expression in the heterozygote, could be made near-recessive, again by selection of the right genetic background. Ford (1955) showed that in the lesser yellow underwing moth, the near-dominance of dark to light in two widely separated natural populations depends on the presence of modifiers specific to the population from which moths were taken. Kettlewell (1965) showed that the dominance of dark to light in the peppered moth collapsed when the English background genotype was replaced with that from a Canadian stock. Clearly, modifiers of dominance are of widespread occurrence and, in some cases at least, are responsible for the presence of dominance as observed in nature.

133

Dominance as a product of evolution

Many years ago, Fisher proposed a general theory of dominance. He suggested that most mutants on first appearance would show some expression when present in heterozygous form. At a later date, however, such mutants might become either recessive or dominant, owing to natural selection for modifiers which either greatly reduced or greatly enhanced the expression of the mutant in the heterozygote. If the mutant were disadvantageous, modifiers reducing the expression in the heterozygote would be at a selective advantage and would spread. Ultimately, the mutant heterozygote would look like the wild-type homozygote, that is, the mutant would have become recessive. Fisher noted that most mutants are recessive at the gross phenotypic level; most mutants are also harmful. He argued that this recessivity had *evolved* via the spread of modifiers. On the other hand, when the mutant was advantageous, modifiers enhancing the expression in the heterozygote towards that of the mutant homozygote would be at an advantage and the mutant would become dominant to the previous wild type.

Ever since this theory was put forward in 1928, it has aroused intense controversy. The arguments have often appeared in mathematical form; for a discussion, see Ewens (1969). It should, however, be made clear that disagreements over details of the mathematics are not the main source of controversy. It is, of course, important to get the mathematics right, since confusion here can only add to difficulties. The main objections to Fisher's theory, however, can be stated quite clearly in words, since they relate to the plausibility of his assumptions. The validity of such assumptions must be tested by observation and experiment (not very easy in practice); note that Ewens appears fully to share this view. We shall, therefore, give a mainly verbal discussion, with only passing references to quantitative investigations.

Disadvantageous mutants

Consider first disadvantageous mutants. At any locus, a specific mutation will appear *de novo* only very occasionally, unless the population be very large, but we can group together all specific mutants having a similar harmful effect (for example, all mutants giving rise to an inactive enzyme), and treat them as the same mutant. Such a mutant will continually rearise and, although often lost at once by chance, will sometimes give rise to descendants which persist for a time before being finally removed from the population by selection and drift. In a very large population, such mutants will be present, at low frequency, most of the time, so that some wild type/mutant heterozygotes are usually present. In these heterozygotes, modifiers making the individual phenotypically more like wild type would be at an advantage; the modifiers are supposed to be neutral in the wild-type homozygote. Hence overall the modifiers have a *small* advantage and spread through the population; the mutant has become recessive.

But (and all are *agreed* on this point) the advantage of the modifiers is very slight, since heterozygotes at the main locus are rare. Thus the whole process takes a very long time (tens of thousands of generations at least). This is where the disagreement arises. If we follow Fisher, we have to suppose that the modifiers remain virtually neutral, in all respects unrelated to their effect on dominance, for all this time. Otherwise the fate of the modifiers will be decided by selection arising from effects at the modifier loci other than those connected with dominance modification, since selection arising from these other effects could be quite small and yet still overwhelm the very feeble selection arising from dominance modification. This point was first raised by Wright, who proposed an alternative theory to account for the recessivity of many unfavourable mutations. Many of these are probably pure loss mutations, giving rise to a near-inactive enzyme or no enzyme at all. If so, the wild type will be dominant to the mutant if the heterozygote produces enough enzyme to meet the requirements of the organism. Thus on this theory, there is no evolution of dominance, the mutant is recessive from the start. Now the evidence available (summarized in Gillespie and Langley 1974) indicates that enzyme activity in heterozygotes is roughly intermediate to activity in the corresponding homozygotes (this in fact seems to be true quite generally, irrespective of whether or not one homozygote is inactive). On Wright's theory, we suppose that about half of the normal activity is sufficient for normal life. It is not very clear why the organism should normally go to the trouble of producing about twice the activity it needs; perhaps the heterozygote, although normal in appearance, is slightly less fit than the wild-type homozygote. Haldane (1939) suggested that natural selection will pick out wild-type alleles giving rise to excess activity, since such alleles will be at a selective advantage in those heterozygotes in which one allele gives rise to no enzyme or near-inactive enzyme and also perhaps in some environments. On Fisher's theory, we must suppose that half the normal activity, while inadequate *per se*, is in some way compensated by the action of the modifiers.

Advantageous mutants

Now consider advantageous mutants. Modifiers *enhancing* the expression of the mutant in the heterozygote are at an advantage and increase in frequency. If the modifiers reach high frequency, the mutant will have become dominant to wild type.

Now, at first, as the advantageous mutant becomes fairly common, there are many mutant/previous wild-type heterozygotes, so that the modifiers have a fairly large advantage and spread fairly rapidly. But, as the mutant becomes very common, most individuals will be mutant homozygotes; there are few heterozygotes at the main locus, and the advantage of the modifiers will have become very much reduced. Numerical calculations indicate that, as expected, the modifiers will in most circumstances become stuck at intermediate frequencies. It is worth noticing that in the case of industrial melanism, dark was (almost completely) dominant to light from the time when melanics began to spread. Remarkably, the modifiers present in English but absent from Canadian stocks themselves show dominance (Kettlewell 1973), English alleles at the modifier loci being dominant to Canadian. Whatever be the explanation for this finding, it is clear that the English modifiers arose for reasons unconnected with industrial melanism.

Balanced polymorphism

If heterozygotes are present at fairly high frequency for a very long time, we have a situation much more favourable for the evolution of dominance than in the cases discussed so far. Thus evolution of dominance should occur most readily in cases of "balanced polymorphism", that is, when selection is such as to maintain a more or less permanent mixture of genotypes at the main locus within a population. Perhaps the case of the lesser yellow underwing moth, mentioned earlier, is an example of this. The best example is that of Batesian mimicry, which we shall discuss in the next chapter.

Metrical characters

In the case of metrical characters, that is, characters showing continuous variation, the difficulties for Fisher's theory which we have discussed earlier may not arise. Previously we have supposed one main locus, and one or more modifier loci. However, individuals within a population differing with respect to a metrical character will often differ at several loci which effect that character. To put it another way, the character is genetically controlled by several loci (of course, the environment also affects the phenotype). This is true generally for metrical characters which have been studied in detail. The same is true for "meristic" characters, such as chaeta number, which vary in units of one but in which the number of phenotypic classes is fairly large. For most purposes, meristic characters may be treated in the same way as metrical characters.

Since a given character is controlled by several main loci, the number of modifier loci could perhaps be considerably *fewer* than the number of main loci. Suppose, then, that the optimum phenotype is at one extreme of the phenotypic range. The intensity of selection at modifier loci could be substantially greater than in most of the cases we considered earlier. This set-up, then, seems very favourable for the evolution of dominance (although it has not, to the author's knowledge, been discussed quantitatively). We should expect to find "unidirectional" dominance; if at any locus, + + represents the genotype of a homozygote giving a phenotype in the direction of the optimum and − − the genotype of a homozygote giving a phenotype in the opposite direction, the heterozygote + − would resemble + + in appearance. We should note, however, that (to some extent) the same outcome could arise without the intervention of modifiers, since alleles at any locus dominant *ab initio*, giving when in heterozygous condition the same fitness as the corresponding homozygote, would be the most likely alleles to get going. Harmful dominants, moreover, would be more readily eliminated than any other alleles.

If the optimum is near the middle of the range, there would be no evolution of dominance; we should expect no dominance or very weak dominance, which would vary in direction from one locus to another ("ambidirectional" dominance). Provided most individuals in the population are near the optimum, alleles dominant *ab initio*, if these exist, would perhaps cause a rather large departure from the optimum and be eliminated.

We conclude that in the case of metrical characters the type of dominance should reflect the type of natural selection operating on the character in the past. With the optimum at one extreme of the range, we should find dominance in the favoured direction; with an intermediate optimum, we should find no dominance or weak ambidirectional dominance. These relationships between present-day dominance and past selection were first pointed out and have been repeatedly emphasized by Mather (e.g. 1960, 1973); there is strong evidence that these relationships hold in practice (e.g. Kearsey and Kojima 1967). It follows (Mather 1960) that the type of natural selection operating in the past can be found from a study of dominance relationships at present. With the occasional exception, present-day natural selection will be the same as in the (not too distant) past, so that the approach via dominance provides a very easily carried out first procedure when one is attempting to study the action of natural selection on metrical traits. For example, in the long-headed poppy, the present author and colleagues (Gale, Solomon, Thomas and Zuberi 1976; Thomas and Gale 1977) inferred, from a study of dominance, that relatively rapid juvenile development would confer a selective advantage. Following this clue, Mackay (personal communication) studied the survival, under adverse winter conditions, of seedlings from autumn germinated 'seed and showed that the more rapid developers did indeed show the greatest survival.

Summary

The degree of dominance of one allele over another at the same locus depends, at least in some cases, on the genetic constitution at other loci. On Fisher's theory, dominance as shown at the present time reflects past selection of the genetic background, disadvantageous mutants having often become recessive, and advantageous mutants dominant; there are some difficulties for this view. Evolution of dominance occurs most readily when a locus is in a state of balanced polymorphism. In the case of metrical characters, natural selection for an extreme optimum should lead to strong unidirectional dominance, whereas selection for an intermediate optimum will give no dominance or weak ambidirectional dominance. Thus the nature of selection in the past can be inferred from the present-day situation in respect of dominance.

CHAPTER TEN

POLYMORPHISM

their weatherings and their marryings and their buryings and their natural selections...

James Joyce, *Finnegans Wake*

Frequency of polymorphism

Before discussing possible explanations for the existence of polymorphism, we outline the main evidence indicating that polymorphism is very common.

Suppose we take a number of individuals from a single population and attempt to select for increase or decrease of a metrical character. If the individuals chosen from the population were genetically identical and homozygous at all loci controlling the character under study, no response to selection would be obtained. If, therefore, we obtain a response to selection, we can conclude that the population must be polymorphic for at least one of the loci controlling the character.

Now, in practice, a response to selection is nearly always obtained, almost irrespective of the character we choose to study. For example, among the characters responding to selection in *Drosophila* (listed in Lewontin 1974) are found body size, wing size, number of abdominal chaetae, number of sternopleural chaetae, development rate, fecundity, egg size, phototaxis, geotaxis, mating preference, expression of a major mutant, sensitivity to changes in environment, and frequency of recombination in a given chromosome region. Failure to obtain a response in *Drosophila* is decidedly unusual, and the same seems to be true for organisms in general. Natural populations must then be polymorphic for some at least of the loci controlling metrical characters.

We can take the matter further and for any given metrical character obtain a minimal estimate of the number of loci controlling that character, for which the population was polymorphic. One way of doing this is to take two lines (derived from the same population), one continually selected for increase of the character, and the other continually selected for decrease of the character, and attempt a rough location of the loci for which the two lines differ (more elaborate versions of the experiment have also been carried out). Whenever experiments of this type have been done, it has become apparent that the population studied must have been polymorphic for a fairly large number of loci controlling the character. For example, from results given by Davies (1971) for *Drosophila melanogaster* derived from a single natural population, it follows that for either abdominal or sternopleural chaetae, several loci were polymorphic on each of chromosomes X, II and III. Moreover, loci controlling abdominal chaetae were distinct from those controlling sternopleural chaetae. This coincides with the usual finding that different metrical characters are, to a considerable extent, controlled by different sets of loci.

Thus all the evidence available from selection experiments indicates that natural populations are polymorphic for a large number of loci controlling metrical characters. On the other hand, the actual *proportion* of such loci which are polymorphic is unknown. Moreover, we do not know to what extent we are dealing with structural loci and to what extent with regulatory loci. Perhaps it is not too hazardous to guess that both types of locus are contributing to the polymorphisms detected.

Turning now to individual loci, we should note that the proportion of individuals heterozygous at a given locus can sometimes be quite large. For example, at several human blood group loci, about 50% of individuals are heterozygous (Race and Sanger 1975). Particularly remarkable is the extent of polymorphism at the four linked loci HLA–A, HLA–B, HLA–C, HLA–D, situated on human chromosome 6, which are critical for the histocompatibility reaction. Among European Caucasoids, for example, 19 alleles with individual frequencies exceeding 1% are present at the HLA–B locus. Over 80% of individuals are heterozygous at any one of the loci HLA–A, HLA–B and HLA–D (Bodmer and Bodmer 1978). However, studies of individual loci *in general* (like studies of metrical characters) cannot give estimates of the extent to which a population is polymorphic, since monomorphic loci are not, *in general*, identifiable. Some human-blood-group antigens, for example, are carried by nearly everyone; the corresponding loci were recognized, of course, only because researchers happened to obtain the appropriate antibodies. We can be almost certain that there are many blood-group loci, showing virtually no variation, which have not been identified.

However, in the case of loci determining the structure of known soluble proteins (in practice mainly enzymes), we can proceed in a much more quantitative manner. For any given population we can obtain minimum estimates of the proportion of loci polymorphic and also of the "average heterozygosity", that is, the proportion of individuals heterozygous at a locus, averaged over loci. Consider any specified polypeptide chain. If individuals within the population differ in respect of amino acid composition at any point in the chain, they must also differ at the locus responsible for the structure of that chain. On the other hand, if all individuals are identical in respect of amino acid composition, this does not necessarily imply identity at the locus, in view of the redundancy in the genetic code. It is, of course, a purely verbal matter whether we regard a locus as polymorphic in cases where individuals differ in respect of DNA composition but not of amino acid composition. Since, however, differences in DNA which are not reflected in amino acid differences are presumably neutral, or almost so, and hence easily explained along the lines given for neutral alleles in earlier chapters, we shall not consider them further. Note, however, that we are also ignoring (perhaps wrongly) any DNA differences in parts of the locus that are not translated. For convenience, then, we define a locus as monomorphic if, apart from rare variants, all individuals in the population have the same polypeptide. Our qualification in regard to rare variants is necessary in order to exclude cases of rare harmful alleles whose presence is attributable to fairly recent mutation. Of course, the precise frequency at which we regard an allele as rare is somewhat arbitrary, but this is not a serious problem in practice. If one allele has a frequency exceeding 0.99, we do not go seriously wrong in regarding the population as monomorphic; this is the criterion normally used.

In practice, following the procedure introduced to population genetics by Harris (1966) and by Lewontin and Hubby (1966), differences between polypeptides are nearly always investigated by gel electrophoresis. For details of the procedure, see e.g. Harris (1975). Differences which do not lead to a difference in electrical charge will not normally be detected by electrophoresis, so that the extent of polymorphism will be underestimated. Nevertheless, studies of this kind revolutionized our view of genetical variation in natural populations; a quite unexpectedly large amount of polymorphism was revealed. In Man, for example, the proportion of loci polymorphic was shown to be 23%, with an average heterozygosity (over all loci studied) of 6.3% (Harris and Hopkinson 1978). In *Drosophila pseudoobscura*, results were even more dramatic; on averaging results for flies from several North American localities, the proportion of loci polymorphic was found to be 43%, average heterozygosity (over all loci studied) 12.8% (Prakash, Lewontin and Hubby 1969; Lewontin 1974). Similar high incidence of polymorphism has been demonstrated in many other species (see e.g. Selander 1976).

As already pointed out, the figures given are certainly underestimates, perhaps quite serious underestimates. Singh, Lewontin and Felton (1976) have listed some of the methods that could be used to discriminate different variants which perform identically on standard electrophoresis. They suggest, as first approaches, varying the conditions of electrophoresis (e.g. pH, gel concentration) and studying loss of activity following standard exposures to heat or urea. For further discrimination, they list study of dimerization potential of different alleles, immunological methods, study of enzyme kinetics, and finally peptide mapping and fingerprinting. In *Drosophila pseudoobscura* collected from 12 localities, they detected, using standard electrophoresis, 8 alleles at the xanthine dehydrogenase locus. By merely varying conditions of electrophoresis, they resolved these 8 alleles into 27; the heterozygosity at this locus (averaged over localities) rose from 44% to 63%. After further resolution by comparing loss in activity due to heat, they concluded that the heterozygosity was at least 72%.

These results are very remarkable but may not be typical. Thus (Coyne and Felton 1977) changing the conditions of electrophoresis in *D. pseudoobscura* at an alcohol dehydrogenase locus raised heterozygosity from 50% to 58%, but at an octanol dehydrogenase locus from 8% to only 10%. Beckenbach and Prakash (1977) failed to detect any new alleles at two hexokinase loci. At any rate, irrespective of the precise amount of polymorphism, there can be no question that polymorphism is an extremely common phenomenon.

Some possible explanations for polymorphism

What are we to make of all this variation? Of course, if most of it represented the action of neutral alleles, the explanation is straightforward. For reasons given in chapters 5 and 6, however, we shall adopt the *working hypothesis* that at least a fairly large portion is due to natural selection. We then have to consider how natural selection could give rise to the observed variability.

Now, according to the view held by most of the early workers in population genetics, there would be, at almost any locus, a homozygous genotype optimal for the population concerned. It was supposed that this genotype remained optimal for a fairly long period of time and that in most populations the optimum had been, near enough, reached as a result of past natural selection. On this view, polymorphism would be unusual.

Since this "classical" view is demonstrably wrong, we seem all set for the argument:

(1) If there is an optimum genotype, nearly all loci would be monomorphic.
(2) Many loci are not monomorphic.
Hence
(3) at many loci, there is no optimum genotype.

Thus we arrive at the possibility that selective advantages vary over space or time, or with the frequency of different genotypes in the population. Now it is perfectly obvious that the conditions of life in many habitats vary from one part of the habitat to another, and from one season or year to another. Moreover, competition between individuals for a given resource will sharpen as the number of individuals attempting to obtain that resource increases. Any genotype conveying a preference for a resource not in popular demand will clearly be at an advantage. This advantage leads to an increase in the frequency of the genotype concerned, giving increased competition for its preferred resource and thus a decline in the selective advantage of that genotype.

It would appear, then, that the right questions to ask are: What proportion of loci are affected by the phenomena just described? By how much do relative selective advantages vary? Is the variation sufficiently large to maintain polymorphism, or do we get monomorphism in spite of the variation?

However, this is not on the whole the way the subject developed. Our premise (1) is not, strictly speaking, correct. We can have an optimum genotype *and* polymorphism, provided the optimum genotype is not a homozygote but a heterozygote. We must, therefore, consider this possibility in some detail.

Heterozygous advantage: theory

Fisher (1922) was the first to point out that heterozygous advantage will (given random mating and two alleles at a locus) necessarily lead to a stable mixture of genotypes. To see this we follow Fisher (1930), as expounded by Edwards (1977).

Let p, q be the frequencies of alleles $\underline{A}, \underline{a}$ in some generation and p^*, q^* the corresponding frequencies one generation later. Given

	AA	Aa	aa
Frequency among zygotes	p^2	$2pq$	q^2
Viability	a	b	c

we find (ignoring drift) that

$$\frac{p^*}{q^*} = \frac{ap^2 + bpq}{bpq + cq^2} = \frac{p}{q}\frac{ap + bq}{bp + cq} = \frac{p}{q}\frac{ap/q + b}{bp/q + c}$$

Write $u = p/q$, $u^* = p^*/q^*$. Then

$$u^* = u\frac{au + b}{bu + c}$$

Now define u_e as equal to

$$\frac{b - c}{b - a}$$

If $\underline{Aa}$ is the most viable of our three genotypes, then b is greater than both a and c; hence u_e is positive and thus a possible value for u. Now

$$u^* = u\frac{au + b}{bu + c} = u\frac{(au + c) + (b - c)}{(au + c) + u(b - a)}$$

$$= u\frac{(au + c) + u_e(b - a)}{(au + c) + u(b - a)}$$

On subtracting u_e from both sides of this equation, we obtain

$$u^* - u_e = (u - u_e)\frac{au + c}{bu + c}$$

If u happened to equal u_e, so that $(u - u_e) = 0$, u^* would also equal u_e and allele frequencies would remain constant. Now, since a is less than b, $(au + c)/(bu + c)$ is less than 1. Hence if u does not equal u_e, $(u^* - u_e)$, while having the same sign as $(u - u_e)$, is smaller than $(u - u_e)$ in magnitude. Thus, irrespective of whether u starts out larger or smaller than u_e, u gradually approaches u_e in successive generations.

We see then that given heterozygote advantage, the population will settle down to a state in which allele frequencies are intermediate; the ratio of allele frequencies p_e, q_e once the population has settled down being

$$\frac{p_e}{q_e} = \frac{b-c}{b-a}$$

whence

$$p_e = \frac{b-c}{2b-c-a}, \quad q_e = \frac{b-a}{2b-c-a}$$

With allele frequencies at these values, a change in frequencies due, say, to a temporary influx of migrants will be automatically corrected, allele frequencies returning to p_e and q_e over successive generations. We are said to have a *stable equilibrium* at frequencies p_e and q_e (note, however, that we have ignored the effects of drift, which turns out in some circumstances to be an awkward complication). It may be shown that, given constant fitness, heterozygote advantage is the only case for which a stable equilibrium exists other than at $p = 0$ or at $p = 1$ (if heterozygotes are at a disadvantage to both homozygotes, the argument analogous to that just given shows that $u^* - u_e$ is greater than $u - u_e$ in magnitude; in other cases, apart from heterozygous advantage, commonsense—or Wright's formula of our chapter 8—shows that p must either always increase to unity or always decrease to zero).

Sickle-cell anaemia

There is only one properly substantiated example of heterozygous advantage involving a single locus, namely sickle-cell anaemia, investigated by Allison (1955). At the risk of boring the reader, who may well be familiar with this example, we outline the evidence for heterozygous advantage in this case, as given by Allison (1964).

Sickle-cell disease is common in childhood in Africa; most cases die in childhood, so that the fitness of the $Hb^S Hb^S$ genotype is near zero. Studies on children in various parts of Africa show that *Plasmodium falciparum*, the cause of subtertian malaria, has a decidedly higher incidence in $Hb^A Hb^A$ homozygotes than in heterozygotes, the ratio of incidences being $1.46:1$. The difference in incidence is even more striking in cases where infection is heavy; there is strong evidence that mortality increases with parasite count (note, however, that these differences in incidence are much less apparent in adults, owing to acquired immunity). Among African children dying of malaria, the proportion of cases with genotype $Hb^A Hb^S$ was very much lower than the local population frequency of that genotype. The frequency of heterozygotes is significantly greater in adults than in infants. Finally, a high frequency of the Hb^S allele is found only in populations living in regions where malaria is, or was until recently, endemic or in population groups, such as Blacks in the U.S.A., whose ancestors came from such regions.

We conclude that there is differential viability in early childhood, with heterozygotes at a selective advantage over either homozygote.

When Blacks were transferred from West Africa to the relatively malaria-free U.S.A., this heterozygous advantage will have been substantially reduced and will have disappeared completely with the elimination of malaria from the U.S.A. in more recent times. On the other hand, Hb^SHb^S remained very disadvantageous throughout. Hence we expect a reduction in the frequency of Hb^S among U.S. Blacks, as compared with the frequency among their West African ancestors, over and above the reduction attributable to race crossing. There is convincing evidence that this extra reduction has in fact occurred (Workman, Blumberg and Cooper 1963).

Is heterozygous advantage common?

Many polymorphisms have been attributed to heterozygous advantage. In most cases, this is merely a guess; certainly no case has been investigated with anything like the thoroughness devoted to sickle-cell anaemia. To suppose, as is sometimes done, that heterozygous advantage is the explanation for most cases of polymorphism, while at the same time taking a rather dismissive attitude to other explanations, is surely unwarranted. Indeed the very assumption that just one type of explanation will account for most cases of polymorphism has not, to date, received any empirical support and may well be unjustified.

The main attraction of heterozygous advantage as an explanation is that, *given the assumptions we listed* in our mathematical treatment, a stable equilibrium is obtained, however small the advantage. The emphasis on stable equilibria may be overdone; we do not know, even roughly, the proportion of polymorphisms which represent stable equilibria. The existence of polymorphism for blood groups, especially Rhesus and Kell, which can give rise to haemolytic disease of the newborn and thus heterozygous *disadvantage*, suggests that stable equilibria may not be as common as is often supposed. But even if this objection is invalid, the argument that even the most minuscule heterozygous advantage gives a stable equilibrium carries rather little weight, since it depends critically on the assumptions of random mating and two alleles at a locus. For species reproducing partly by selfing, partly by random mating, it may be shown (Hayman 1953, Workman and Jain 1966) that as the amount of selfing increases, increasing heterozygous advantage is required for a stable equilibrium to be obtained. This provides a check, although not a rigorous one, on the view that heterozygous advantage is the main source of polymorphism. If this view were correct, we should rather expect species showing a high degree of selfing to be very much less polymorphic than outbreeders; this expectation is not borne out in practice, either for metrical characters or proteins (Allard, Jain and Workman 1968; Marshall and Allard 1970; Allard and Kahler 1971). Moreover, comparisons between populations within highly inbreeding species, and also between such species, show no clear relationship between mating system and amount of genetic variability (Allard, Kahler and Clegg 1975).

Further, even if random mating is assumed, there are severe restrictions on the number of alleles that can be maintained in stable equilibrium by heterozygous advantage (Kimura 1956, Mandel 1959). As Lewontin, Ginzburg and Tuljapurkar (1978) have shown, very special values of fitnesses would be required for more than six alleles to be maintained at a single locus, even in cases where *all* heterozygotes are fitter than *all* homozygotes. They have, indeed, identified a range of values which would give stable equilibria with many alleles present, but in such cases allele frequencies would come out almost equal. In fact, rather marked *inequality* of frequencies is the norm, at least for variation detected electrophoretically. They conclude that heterozygous advantage cannot account for loci where there are a large number of alleles within a population, most alleles being rather rare, as is the case for xanthine dehydrogenase in *Drosophila* and placental alkaline phosphatase in Man.

Finally, if heterozygous advantage were the main cause of polymorphism, we should expect little polymorphism in haploids. Yet extensive studies on *Aspergillus nidulans* (summarized in Croft and Jinks 1977) have revealed a large amount of polymorphism (heterokaryons are very infrequent in nature and are thus not a complication). Similar results are found in *Aspergillus amstelodami*, *Neurospora crassa* and *Ceratocystis ulmi* (reviewed by Caten 1979) and also in *Escherichia coli* (Milkman 1975).

It is clear, then, that there are many polymorphisms which either cannot be, or are very unlikely to be, explained by heterozygous advantage. This does not, of course, exclude the possibility that heterozygous advantage is still important in a fair proportion of cases, but no convincing evidence is available to support this idea. We shall therefore turn now to some other situations which can give rise to polymorphism.

Variable selective advantage: formal theory

For a theoretical treatment of variable selective advantages we can, in principle, proceed as we did for heterozygous advantage. We assign viabilities (algebraically) to the different genotypes under different conditions and see what happens (in particular, whether we reach a "stable polymorphism", that is, a stable equilibrium with two or more alleles present).

Unfortunately, this approach is not very helpful. Consider, for example, cases where selective advantage varies over space or time (reviewed by Felsenstein 1976). We can obtain conditions which viabilities have to obey to give a stable polymorphism, but these conditions look bizarre and "unnatural". If we suppose that, in nature, viabilities can vary in a virtually unrestricted manner (over a given range of values), we shall conclude that viabilities which obey the conditions for stable polymorphism, being only a small fraction of possible viabilities, are rarely encountered in practice. But this supposition is surely unjustifiable; there are, presumably, unknown rules which govern the variation in viabilities. As Gillespie and Langley (1974) point out, the enzyme activity of a heterozygote is, quite generally, about half-way between the activities of the corresponding homozygotes; this will impose a fairly severe restraint on variation in viabilities. If we suppose further that fitness increases with increasing enzyme activity, the curve gradually flattening out as activity rises, we find that the conditions for stable polymorphism, with many alleles present, are rather easily met (Gillespie 1977). Even if this last assumption (for which there is no critical evidence for or against) turns out to be wrong, the point is still made that the imposition of plausible restraints on variation in viabilities makes a great deal of difference to our conclusions.

Similar difficulties apply to frequency-dependent selection (in the usual restricted sense), that is, when the selective advantage of any genotype wanes as the frequency of that genotype rises. The precise relationship (for a given genotype) between fitness and frequency may, perhaps, be critical to the outcome, but we lack the data which would give us appropriate formulae describing this relationship. In practice, formulae are chosen on the basis of simplicity, not experience, so that the relevance of the results obtained is uncertain. As far as they go, results (Clarke and O'Donald 1964, Wright 1969) do confirm the widely held intuitive view that very mild frequency-dependent selection is sufficient for the maintenance of stable polymorphism.

Finally, we should note that the relative viabilities of different genotypes at a locus will usually depend on the particular alleles present at *other* loci. We should, therefore, consider two or preferably more loci simultaneously. When this is done for constant viabilities (see especially Karlin 1975, Karlin and Carmelli 1975) it emerges that the outcome in any particular case depends critically both on the viabilities and on the degree of recombination between the loci concerned; no doubt the same is true for variable viabilities. Extreme caution, then, is essential if one is attempting to infer possibilities for stable polymorphism from a theoretical treatment of one locus at a time.

Measuring selective advantage: formal approach

We have argued that theory is, at the moment, of limited help to an investigator of the causes of polymorphism. Why not, then, go and measure the fitnesses found in practice? We could, it would appear, make substantial progress by taking a purely "formal" approach, that is, measuring differences in fitness without regard to the biological source of these differences. However, a *very large* number of individuals will have to be scored if we are to estimate differences in fitness with any accuracy.

Suppose, for example, we wish to compare the viabilities of two genotypes $\underline{A}, \underline{B}$ of *Drosophila*. For simplicity, we consider a laboratory experiment, although the same considerations (with many added practical difficulties) would apply to a study in the natural habitat. We start with an equal number of eggs of each genotype and count the number of adults of each type emerging. Suppose the unknown relative viabilities of $\underline{A}, \underline{B}$ are $(1+s):1$. We can obviously estimate s from the number of adults of each type emerging. If x adults of type $\underline{A}$ and y adults of type $\underline{B}$ emerge, our estimate of s is

$$\hat{s} = \frac{x}{y} - 1$$

The estimated standard error of $\hat{s}$ proves to be

$$\sigma(\hat{s}) = (2+\hat{s})\sqrt{\frac{1+\hat{s}}{N}}$$

where $N \ (=x+y)$ is the total number of adults emerging. Although s is unknown, we can (from standard statistical theory) be confident that s lies (near enough) between the limits

$$\hat{s} - 1.96\sigma(\hat{s}) \quad \text{and} \quad \hat{s} + 1.96\sigma(\hat{s})$$

Suppose we raise 10 000 adults and find $\hat{s} = 0.05$. We arrive at the singularly unhelpful result:

We can be confident that s lies between 0.0088 and 0.0912.

Had we raised 40 000 adults and found $\hat{s} = 0.05$, we should have done rather better, obtaining limits 0.0294, 0.0706; while with 160 000 adults, we should obtain the much more helpful limits 0.0397, 0.0603. We recommend the reader to try various values of N and $\hat{s}$. For given N, the ratio of upper limit : lower limit falls as $\hat{s}$ rises; moreover, when selection is intense, it will not be so critically necessary to have really accurate estimates. Where selective advantages are less than 0.1, however, the whole approach seems out of the question, even for a fairly large team of investigators, since we should obviously like to repeat the experiment under different sets of conditions.

Even if we are content merely to detect a difference in fitness and not to estimate it, the same problem appears in a less acute but still unnerving form. In our experiment, a difference in viability is demonstrated if the number of A adults emerging: the number of B adults emerging departs significantly from 1 : 1, as shown by the usual χ^2 test. Suppose that, if there is a viability difference, we wish to be able to detect this difference (at the 5 % significance level) in 95 % of experiments of the type under discussion. It may be shown that we must raise at least

$$N = \frac{12.995(2 + s)^2}{s^2}$$

flies in all. Thus if $s = 0.05$, we must raise a minimum of 21 845 adults, although much smaller numbers will do if s is large. Similar difficulties arise in all other types of experiment for which the calculations have been done. For further discussion, see Lewontin (1974).

Finally, we should note that even when differences in fitness have been determined with reasonable accuracy, their interpretation is not always easy; predictions can go badly wrong. Suppose we find, experimentally, the fitness of genotypes at a given locus and attempt to predict, from these fitnesses, future changes in allele frequency at that locus. The changes that actually occur may deviate markedly from expectation if, unknown to us, the relative fitnesses of our genotypes vary with allele frequencies at *other* loci. The reader is referred to Lewontin (1974, chapter 6) for further discussion of this problem.

The ecological genetic approach

The difficulties we have described are formidable, but become less serious if we use our knowledge of the population biology of the organism under study. Take as example visual selection in *Cepaea nemoralis*. Differences in fitness between the different morphs were disclosed very readily, because Cain and Sheppard studied frequencies of morphs, not just in the original populations, but also among individuals *that had been predated*. By taking account of the fact that thrushes eat snails, a substantial advance in understanding was achieved. Moreover, one could have some confidence that the detected differences in fitness were meaningful in their own right, because the explanation in terms of cryptic colouration makes sense. We are not, of course, claiming that it is easy to discover the role of a given character in the life of an organism. Indeed, we indicated some difficulties when discussing *Cepaea* in chapter 5; comparable difficulties have appeared in other cases, such as wing spotting in *Maniola jurtina* (e.g. Dowdeswell 1975) and cyanogenesis in *Lotus corniculatus* (e.g. Jones 1972; Ellis, Keymer and Jones 1976, 1977). Rather, we suggest that to attempt biological explanations and then try to obtain evidence relevant to these explanations is the most useful approach at present. Although this view is fairly widely held, some consider that an appropriate modification of the formal approach would be effective (see e.g. Lewontin 1974, chapter 6).

We illustrate these remarks by describing, in outline, two well-known studies of polymorphism (for a much fuller discussion see Dobzhansky 1970 and Ford 1975).

Inversions in *Drosophila pseudoobscura*

In chapter 5 we described the seasonal fluctuations in frequency of different chromosome arrangements in *Drosophila pseudoobscura*. We noted that frequencies in a given locality changed during the year, and that the same pattern of change was found in successive years, although this pattern varied with locality. We turn now to the interpretation of these results.

In parallel with their observations in nature, Dobzhansky and his colleagues set up laboratory population cages, in which changes in frequency under defined conditions could be observed. Relative fitnesses of the different arrangements were found to depend on temperature, type of food and degree of crowding. In addition, the relative fitness of two arrangements is affected by the presence of other arrangements.

When cages were set up with flies all drawn from the same locality, it was often found that individuals heterozygous for a pair of arrangements were fitter than the corresponding homozygotes (this has no bearing on our earlier discussion on possibilities for heterozygous advantage at a single locus, since chromosome arrangements differ at many loci). When, however, arrangements differ in place of origin, heterozygous advantage is unusual. For example, arrangements ST and CH in combination gave heterozygous advantage (at 25°C but not, interestingly, at 16°C) only when originating from the same locality. Thus visually similar arrangements can behave differently in cages when differing in origin, either because of differences in their genetic content or because of interaction with loci outside the inversion which differ with locality. This helps to explain why the pattern of fluctuation varies with locality.

The reader will recall that at Piñon Flats, the frequency of ST fell sharply from March to June and rose sharply from June to August, while the reverse was true of CH. This pattern can be imitated in cage experiments. In spring, following winter elimination, numbers will be relatively low and larval competition for food minimal. In cages where larvae were supplied with abundant food, CH/CH proved superior in viability to ST/ST, so that CH rose in frequency. For the hot summer months, with intense competition, conditions can be approximated by cages with limited food supply maintained at 25°C. Under these cage conditions, heterozygous advantage, with ST/ST also fitter than CH/CH, comes into play, leading towards a stable equilibrium with an excess of ST; thus the frequency of ST rises. However, patterns of change in some other populations are not so easily imitated.

These studies provide an excellent example of seasonal fluctuations in selective advantage, in which some of the relevant environmental variables have been identified. Nevertheless, substantial problems remain. As Dobzhansky (1971) remarked:

> The nature of the selective forces, the results of the operation of which are so plainly visible, is still conjectural—despite more than thirty years of study.

We suggest that the source of the difficulty is as follows. Very many loci are contained within any inverted region; presumably the unknown products of these loci interact in a complicated unknown manner. Hence we have no idea why the inversions react to environmental variables in the way that they do. While, then, we can follow the changes that have occurred, we have no guide to identifying relevant environmental variables, save in the case of cyclical selection, where we can attempt to relate repeated seasonal changes in frequencies to recognizable seasonal changes in the environment. Even here we are proceeding empirically, so that many relevant factors are difficult to identify. Predicting the effects of a given *long-term* change in environment is thus very hazardous. Moreover, observed long-term changes in frequencies will often be difficult to explain. Over the period (roughly) 1940–1970, for example, the frequency of the arrangement Treeline (TL) rose markedly over a very wide area of the Pacific Coast (Anderson, Dobzhansky, Pavlovsky, Powell and Yardley 1975). A satisfactory explanation for this change has not been found.

Thus, despite the fact that selective differences are large, that heterozygous advantage has been found, and that interactions between inversions and other loci can be studied, many puzzles remain. Although the set-up seems ideal for a formal approach, the predictive power of this latter approach seems very limited even here. On the other hand, the "ecological genetic" approach, although in some ways very successful, has been handicapped by our rather fragmentary knowledge of the ecology of *D. pseudoobscura*, coupled with the extreme complexity of the genetic system locked up in any inversion. It is helpful, therefore, to consider a case where the relevant ecology is well understood, and the genetic system not too complicated; in such a case, the virtues of the ecological genetic approach, particularly in regard to prediction, should be readily apparent. We take as example Batesian mimicry; for simplicity we confine our attention to butterflies (for an extensive discussion, see Turner 1977).

Batesian mimicry in butterflies

In the type of mimicry known as Batesian, the model is distasteful to predators (in the case of butterflies the chief predators are birds) while the mimic, which is palatable to predators, is protected from attack by its resemblance to the model; the resemblance is particularly striking when the butterflies are in flight. It is clear from laboratory experiments that the predators *learn* to associate a particular pattern of appearance with distastefulness from exposure to the model. Predators which have not been exposed to the model eat the mimic, but after repeated exposure to the model reject the mimic. For the resemblance to the model to benefit the mimic, then, the predator must encounter the model much more frequently than the mimic, since otherwise the predator will learn to associate the pattern of appearance with tastefulness. Thus Batesian mimicry is an example of frequency-dependent selection; as long as the mimic is rare compared to the model, it has a selective advantage over non-mimetic forms, but this advantage wanes as the frequency of the mimic rises. Consider then a case where a particular mimic is fairly common. If a mimic of a different model arises by mutation, it will initially be rare and therefore at an advantage over the old mimic. Hence it will increase in frequency, with an accompanying fall in fitness, until eventually both mimics are equally fit. Thus ultimately two or more mimetic forms belonging to the same species coexist indefinitely; we have a stable equilibrium. Indeed, non-mimetic forms may coexist with mimetic forms of the same species, since the mimetic forms, which are conspicuous, may be less fit when at high frequency than non-mimetics.

Thus, from the theory of Batesian mimicry we can make the following predictions. Firstly, that the geographical distribution of different mimetic forms within a given species should reflect the geographical distribution of the corresponding models; this is generally the case in practice. Secondly, polymorphism should be very common; this is indeed so. The reader is referred to Ford (1975) for a detailed account of the very elaborate pattern of genetical variation in the swallow-tail butterfly *Papilio dardanus*, most of which, as Ford shows, is easily explained on the principles we have described.

The most interesting predictions concern the dominance relationships of the various alleles controlling the mimetic patterns. Take as example *Papilio dardanus*; major differences between the morphs represent differences at a very closely linked group of loci, acting effectively as a single locus, while more minor differences represent variation at a number of other loci. Consider, then, a set of alleles, at the various loci, which jointly give rise to a particular mimetic pattern. It is very unlikely that, in the absence of dominance, individuals heterozygous for these alleles and individuals homozygous for these alleles will be equally good mimics. If the homozygote is the best mimic, we expect natural selection for modifiers which make the heterozygote look like the homozygote, that is, evolution of dominance in the usual sense. If the heterozygote is the best mimic, we expect selection of modifiers making the homozygote look like the heterozygote; in this case also we end up with dominance.

Generally, then, we expect dominance whenever there has been occasion for the selection of the appropriate modifiers but not, of course, otherwise. Thus we should obtain dominance in the progeny of crosses between forms that coexist, but not between forms that live in separate localities. That this is usually the case in practice has been shown by Clarke and Sheppard in an extensive series of investigations (1960, 1971, 1972) on *Papilio dardanus*, *Papilio memnon* and *Papilio polytes*.

Final comments

We have not, of course, discussed all the examples of polymorphism which have been studied in detail, although we have at least mentioned a fair proportion of them in various places in this book. Unfortunately, these studies are very time-consuming. For the moment, the number of detailed investigations is too small for any generalizations to be made with confidence, except perhaps that it does not seem too difficult to find examples of any of the mechanisms proposed as causes of polymorphism (this applies also to some other mechanisms, such as "meiotic drive", which we have not discussed, as probably being important only in rather special cases). On the other hand, one should not under-rate the importance of the studies that have been carried out. At the very least, they show that the theories and ideas of population genetics, derived in the first instance as plausible consequences of the Mendelian mechanism, do have relevance in natural populations.

Summary

Polymorphism is very widespread. It has often been suggested that most cases of polymorphism are due to heterozygous advantage. This is unlikely to be correct, since communities of inbreeding diploids and of haploids are about as variable genetically as communities of outbreeding diploids. Variation of selective advantage over space or time, or with genotypic frequency, may all be important in the maintenance of polymorphism. There can indeed be no doubt of their importance in some individual cases studied in detail, but the data are insufficient for general conclusions to be drawn.

Theoretical analysis of polymorphism is hampered by our limited knowledge of actual selective values and of their variation in nature. Very large-scale experiments are required if these selective values are to be measured with accuracy. This difficulty can, to some extent, be circumvented in cases where the biological role of the character under investigation is understood.

CHAPTER ELEVEN

EVOLUTION OF ALTRUISM

In the case of social insects there is no limit to the devotion and self-sacrifice which may be of biological advantage in a neuter. In a beehive the workers and young queens are samples of the same set of genotypes, so any form of behaviour in the former (however suicidal it may be) which is of advantage to the hive will promote the survival of the latter, and thus tend to spread through the species.

J. B. S. Haldane, *The Causes of Evolution* (1932)

Altruistic behaviour

Hitherto we have considered natural selection as favouring some individuals within a population (namely, those individuals possessing a genotype giving high viability or fertility) at the expense of others within the same population. This is known as "individual selection".

How are we to explain "altruistic" behaviour, that is, behaviour which promotes the survival, not of the individual concerned, but of other individuals? Such behaviour, indeed, will often reduce the chance of survival of the altruist.

For example, a bird warns others of the approach of a predator; while his alarm call helps the other members of the flock, it draws attention to himself, thus increasing the chance that the predator will attack him. As a second example, we note that unattached Adélie penguins help defend nests and crèches of chicks belonging to other penguins against attacks of skuas. Many other examples are given in Wilson (1975). We shall discuss possible explanations for altruistic behaviour, with the understanding that at the moment such explanations should, in the present author's view, be regarded as rather tentative. For a recent review, see Parker (1978).

Kin selection

Some altruistic behaviour can be explained on the following grounds: natural selection will favour those genotypes which give individuals who

156

are most efficient at ensuring the ultimate survival of genes identical in structure with their own genes. Such survival can be aided in several ways: firstly, by behaviour which enhances the chance of survival of the individual concerned; insofar as such behaviour is selected, this is just a special case of the usual selection for alleles conveying high viability; secondly, by transmitting one's genes to one's offspring and behaving in such a way as to increase their chance of survival. Selection in this case is for high fertility and capacity to acquire a mate, plus the altruistic character, high parental care of the young.

So far, the points we have made are obvious enough. However, consider now a third pattern of behaviour: the individual behaves in such a way as to increase the probability of survival of his sibs. Any individual has on average half of his genes in common with one of his sibs (chosen at random). Hence if I save my baby sister from drowning, I am helping to ensure the survival of genes identical in structure with my own (this must be so; it does not necessarily follow that this provides a total or partial explanation for my behaviour).

Generally, alleles giving rise to characteristics which enhance the survival of close relatives will be favoured. This is known as "kin selection" (Maynard Smith 1964) and should lead to the evolution of altruistic attitudes towards close relatives. However, such action will often carry risks to the individual concerned. If so, the outcome will be natural selection for an optimum strategy; this may involve both selfish and altruistic behaviour, but will on balance maximize the ultimate survival probability of genes identical in structure with the individual's own genes. The importance of this phenomenon was first indicated by Fisher (1930) and by Haldane (1932a, 1955); the subject was discussed extensively by Hamilton (1964) and more recently by Wilson (1975) and Dawkins (1976). It would be interesting to see whether altruistic behaviour of this kind can be produced or increased by artificial selection. Alternatively, one can look at cases of extreme altruism in nature and see if there are any features of the species concerned which would be particularly conducive to kin selection. The social insects would seem to provide an excellent example.

Social insects

Colonial life in the insects has arisen independently at least eleven times in the Hymenoptera (at least twice in wasps, at least eight times in bees, and at least once in ants) but only once in any other order of insects, the termites (Isoptera).

Now all Hymenoptera (but very few other insects) have an unusual method of sex-determination, "haplodiploidy"; unfertilized eggs give males, which are haploid, fertilized eggs give females which are diploid.

While the details of colonial life vary from one colonial species to another, we can take as example the honey bee *Apis mellifera*. The queen (fertile female, diploid) lays both unfertilized and fertilized eggs. An unfertilized egg develops into a drone (fertile male, haploid). A fertilized egg develops into a queen, provided the larva has been given a continued diet of royal jelly. Otherwise, a fertilized egg develops into a worker (sterile female, diploid). The queen produces a pheromone "queen substance" which inhibits ovarian development in the workers.

Some species differ in detail from this pattern. In some species, workers lay unfertilized eggs to give haploid males, in ants there is a soldier as well as a worker caste, but all species of social Hymenoptera share the following features: firstly, extreme altruism—the workers carry out all the work of the colony for no apparent advantage; secondly, these altruists are always female, the males do no work apart from reproduction.

Hamilton pointed out that the evolution of colonial life in the Hymenoptera was greatly aided by haplodiploidy. Consider two sisters. They have all of their father's alleles in common plus, on average, one-half of their mother's alleles; on average, then they have three-quarters of their alleles in common. On the other hand, mother and daughter have only one-half of their alleles in common. Thus a female is more likely to ensure the survival of alleles identical with her own if she behaves in such a way as to help a fertile sister than if she has a daughter and cares for that daughter. To put the matter another way, consider any mutant allele giving rise to a tendency to care for a fertile sister rather than to have a daughter and care for the latter. Such a mutant is bringing about behaviour more likely to ensure its own survival and transmission than would be the case for an allele giving "normal" female behaviour. Thus the mutant is at a selective advantage, provided it is not expressed in the queen; as Haldane (1932a) pointed out, alleles causing unduly altruistic behaviour in the queen will be eliminated. Some variation between females in fertility, however brought about, would seem therefore to be a prerequisite for the whole system to get started.

Now consider the drone. If we write his mother's genotype at a given locus as G_1G_2, we see that, since half her male progeny are G_1 and half

G_2, there is only a half chance that his alleles are present in one of his brothers, randomly chosen. The same applies to his sisters. On the other hand, all of his alleles are present in every one of his offspring (all of which are daughters). Thus we expect the behaviour of the drones to be the reverse of that of the workers; mutants which when expressed in the haploid lead to a concentration of resources on reproduction would be at a selective advantage. Note that in the termites, which do not have haplodiploidy, the males do as much work as the females. Those interested in pursuing further the whole problem of kin selection in social insects should consult Trivers and Hare (1976) and the references we gave earlier.

Group selection

It is difficult to say how much of altruistic behaviour in general is due to kin selection. Another possibility is as follows. Suppose that the species is made up of a number of sub-populations which rarely interbreed. In some sub-populations, individuals are all altruistic; in other sub-populations, all are selfish. Those sub-populations consisting of altruistic individuals are more likely to survive than sub-populations made up of selfish individuals; thus natural selection picks out sub-populations made up of altruistic individuals. Thus we have selection, not between individuals but between groups of individuals—"group selection".

But (Maynard Smith 1964, Williams 1966) there is a serious difficulty for the argument: how can a sub-population of altruistic individuals arise in the first place? Clearly, not by natural selection, therefore only by drift. But such a sub-population could not remain full of altruistic individuals for long, because a mutation conveying selfish behaviour will be advantageous and sweep through the sub-population. The same happens if selfish behaviour arises, not through mutation, but by migration from a sub-population of selfish individuals.

The only countervailing factor to this is the reduced survival rate of sub-populations containing selfish individuals. It may be shown (Levins 1970; Boorman and Levitt 1973) that for this type of group selection to lead to evolution of altruistic characters, a virtually impossibly high rate of extinction of sub-populations containing selfish individuals must occur.

It seems unlikely, then, that group selection is responsible for most cases of altruism present in natural populations. However, it would be premature to say that group selection is never of biological importance.

It has, indeed, been suggested that the maintenance of sexual reproduction, the biological advantages of which are *perhaps* mainly long-term, is due to group selection. However, one can postulate short-term advantages, so that individual selection may be a major factor. The subject is, unfortunately, too intricate to discuss here; for a detailed account, see Maynard Smith (1978).

Summary

Altruistic behaviour is fairly common. In general, such behaviour will spread under natural selection if it benefits close relatives of altruists, while not doing too much harm to the altruists themselves. In the case of the Hymenoptera, the method of sex determination has facilitated the evolution of extreme altruism manifested in colonial species. In principle, altruism could evolve as a result of enhanced survival of groups of altruists, as compared with groups of selfish individuals, but the conditions under which this could occur are unlikely to be realized at all often in practice.

CHAPTER TWELVE

SUMMARY. SOME TENTATIVE CONCLUSIONS

Biologists work very close to the frontier between bewilderment and understanding.

Sir Peter Medawar,
Induction and Intuition in Scientific Thought (1969)

We set out in the hope of understanding the "genetics of the evolutionary process" (Dobzhansky 1970). We have obviously got somewhere; nevertheless the reader may feel a sense of puzzlement. To what extent have we achieved our objective? G. H. Hardy, to whom our subject is so much indebted for a clear start, was accustomed to award marks in recognition of a wide variety of human qualities, admirable or otherwise, including intellectual achievement. How many out of ten might he fairly have given in our case?

It will be apparent from our earlier discussions that theoretical population genetics has reached an advanced stage of development. For quite a range of different situations, we can predict what will happen and how long it will take for the various changes to come about; in cases where drift is important, we can formulate our predictions in terms of probabilities. Moreover, these theoretical studies have yielded results of very great importance to us. For example, if we wish to talk sense about the evolution of natural populations, we must constantly bear in mind that a very small selective advantage, or a very low rate of migration, can have a critical effect on what happens. Yet we do not quite know what to make of the theoretical findings. Thus, while we can scarcely doubt the dominating role of natural selection for loci controlling characters of obvious adaptive significance, we are uncertain of the role of natural selection in other cases. We have three schemes: the neutralist, mild selectionist, and strong selectionist, with no agreement as to which scheme is, in general, appropriate. The problem is that all three schemes are consistent with Mendelian inheritance; clearly we shall get nowhere if we use such consistency as our sole guide. No doubt we can all agree with Kempthorne (1957) as to "the mathematical beauty of the Mendelian mechanism". Our problem, however, is to decide between three godesses of equal beauty and charm. To do this, we must turn to observation and experiment—"the contemplation of brute fact" (Whitehead). The same is true, of course, in other cases where theory does not give a decisive answer, such as the evolution of dominance or the importance of hitch-hiking. Of course, much excellent observational and experimental work has been carried out. Nevertheless, it would probably be generally agreed that the main factor holding up progress at the moment is the lack of appropriate data. Our problem is not a shortage of explanations for a given set of phenomena, but rather an excess of plausible explanations, with no means as yet of deciding between them (Maynard Smith 1977).

The collection of such relevant data is often difficult. Even so, in the author's opinion there are grounds for guarded optimism. The detailed investigations on loci controlling morphological characters have usually revealed the action of natural selection, often at an unexpectedly large intensity. There seems no reason why similar studies could not be carried out on loci controlling enzyme structure and activity; here the work on alcohol dehydrogenase and on amylase seems very promising. If natural selection proves to be the critical factor at most loci, we still have to explain polymorphism. Here we are undoubtedly handicapped by the practical difficulty of obtaining accurate estimates of relative fitness. Nevertheless, even a qualitative picture of the selective forces acting at a set of loci would be helpful. After all, this is all we have, even in the case of sickle-cell anaemia, where we have achieved a clear understanding of what is happening in spite of the fact that *reliable* quantitative estimates of relative fitness do not exist (estimates of the selective advantage of the heterozygote, although possibly correct, depend on the questionable assumption that there is a stable equilibrium at the haemoglobin locus in the African populations studied).

It may turn out that extended studies will fail to reveal natural selection at a fair proportion of loci. If this happens, the case for the neutral theory will be very much strengthened, although it might still be argued that small selective effects were still present at these loci and could have been detected had we carried out an even more extensive investigation. Indeed, the detection of very small selective differences might justifiably be described as an "extremely intractable problem", analogous to those mathematical problems, soluble in principle, whose solution would require an ideal computer to operate for a period comparable to the age of the universe (Stockmeyer and Chandra 1979); we might describe a genetical problem as extremely intractable if its solution requires us to score as many individuals as the world population of the species. However, the neutralist need not rely on failures to detect selection. The most unexpected prediction of the neutral theory is the constancy of the neutral mutation rate per year, for the same locus in different species. Although the neutral mutation rate cannot be measured directly (since we should have to identify which mutants were neutral) we can at least say that, provided the fraction of mutants at a given locus which are neutral is roughly the same in different species, the overall mutation rate per year at the locus should be about the same in different species. An experimental test of this prediction, although very time-consuming, is certainly possible.

In the absence of such definitive studies, what can we say in summary of our discussions? The author can only give his own (tentative) view, which would not be shared by all. The conditions under which the neutralist scheme would work seem very restricted. Moreover, the observational and experimental studies that have been done, although limited in scope by the time required to obtain reliable results, strongly suggest that rather large differences in fitness are very common. It would appear, then, that natural selection is the prime mover of the process of evolution. If this is so, it seems rather likely that relative fitnesses will undergo fairly marked fluctuations as the habitat varies, and that these fluctuations are the main reason for polymorphism. Perhaps the reader will disagree with these conclusions. All the more reason for him to obtain data in support of his view! Indeed, this book was written in the hope of persuading more biologists to work on the intriguing problems we have described. We conclude with the words of Shakespeare:

> A great while ago the world began,
> With hey-ho, the wind and the rain;
> But that's all one, our play is done,
> And we'll strive to please you every day.

(Twelfth Night)

A MISAPPLICATION OF THE HARDY–WEINBERG LAW

Now, reader, I have told my dream to thee,
See if thou canst interpret it to me,
Or to thyself, or neighbour; but take heed
Of misinterpreting, for that, instead
Of doing good, will but thyself abuse;
By misinterpreting, evil ensues.

John Bunyan, *The Pilgrim's Progress*

Problem

A university lecturer sets his class an exercise on the Hardy–Weinberg law. In making up this exercise, he supposes *random mating and selection in favour of the* AA *genotype*. For example, he may suppose that among the newly formed zygotes frequencies are AA 0.25, Aa 0.50, aa 0.25, and that relative viabilities are AA 1, Aa 0.8, aa 0.4. Finally, he supposes that a random sample of 3000 adults is taken from the population after selection has acted. Since the relative frequencies of the three genotypes in the population after selection will be AA 0.25, Aa 0.40, aa 0.10 (absolute frequencies being obtained by dividing relative frequencies by $0.25 + 0.40 + 0.10 = 0.75$), he tells the class that his sample was as follows:

	AA	Aa	aa	*Total*
Number observed	1000	1600	400	3000

He gives the class no other information apart from these figures. The situation of the class, then, is precisely analogous to that of an investigator whose sole knowledge of a population consists of the numbers of the three genotypes in a sample from that population. He asks the class to comment on these figures.

Next day, the whole class declares that there is a significant departure from Hardy–Weinberg; however, they have failed to agree on the reasons for this departure. A majority attribute the departure to *heterozygous advantage* or possibly to *anti-assortative mating*. Asked to justify their opinions, they point out that they have all found

	AA	Aa	aa	Total
Number observed	1000	1600	400	3000
Number expected	1080	1440	480	3000

The departure of observed from expected is highly significant ($\chi^2 = 37.0$, d.f. = 1, $P < 0.001$) with an excess of heterozygotes and a deficiency of homozygotes. These calculations are indeed correct. Yet the minority of the class who questioned the conclusion of heterozygous advantage or anti-assortative mating were justified in their scepticism. What has gone wrong?

Although this difficulty was expounded by Lewontin and Cockerham (1959) it is not always realized and results of the kind obtained by our class, when arising from real data, have sometimes been regarded as evidence for heterozygote advantage (in cases where anti-assortative mating is considered unlikely). The problem, then, is important in practice. We shall, therefore, stay with our lecturer as he sketches the solution to the paradox.

Solution

Suppose we have

	AA	Aa	aa
Frequency among newly formed zygotes	p^2	$2pq$	q^2
Viability	a	b	c
Frequency after selection has acted	$\dfrac{ap^2}{\bar{w}}$	$\dfrac{2bpq}{\bar{w}}$	$\dfrac{cq^2}{\bar{w}}$

where $\bar{w} = ap^2 + 2bpq + cq^2$ (see chapter 8). If only we could estimate p and q, all would be well. But we can't! We certainly can estimate allele frequencies *after selection* but these are not p, q but

$$p^* = \frac{p}{\bar{w}}(ap + bq) \quad \text{and} \quad q^* = \frac{q}{\bar{w}}(bp + cq)$$

(see chapter 8). If then we take these allele frequencies and calculate proportions "expected", we are calculating, not p^2, $2pq$, q^2 but p^{*2},

$2p^*q^*$, q^{*2}; on subtracting these from the observed proportions $ap^2/\bar{w}$, $2bpq/\bar{w}$, $cq^2/\bar{w}$, we find, after some simple algebra, the bizarre result

Genotype	Proportion observed–Proportion expected
AA	$\dfrac{p^2q^2}{\bar{w}^2}(ac-b^2)$
aa	$\dfrac{p^2q^2}{\bar{w}^2}(ac-b^2)$
Aa	$\dfrac{2p^2q^2}{\bar{w}^2}(b^2-ac)$

We see then that an excess of <u>Aa</u> and a deficiency of both <u>AA</u> and <u>aa</u> will occur if

$$b^2 > ac$$

Now by heterozygous advantage we mean $b > a$ and $b > c$. If heterozygous advantage is present, b^2 will certainly exceed ac, but *the converse does not necessarily hold*. Thus in our case

$$a = 1, \quad b = 0.8, \quad c = 0.4$$

so that b does not exceed a; nevertheless $b^2 = 0.64$, $ac = 0.4$, so that b^2 does exceed ac.

General comment

While one may justifiably test for agreement with the Hardy–Weinberg law in the standard way given above, the direction of departure of observed from expected values cannot be used to infer the type of selection acting at the locus concerned, even if we have reason to believe that selection is the cause of the departure. For other difficulties in this general area, see Lewontin and Cockerham (1959).

APPENDIX 2

SOME POINTS IN STATISTICS

The doctrine of chance...is, in its essence, purely mathematical; and thus we have the anomaly of the most rigidly exact in science applied to the shadow and spirituality of the most intangible in speculation.

Edgar Allan Poe, *The Mystery of Marie Rogêt*

Probability

We do not attempt mathematical rigour, but rather hope to give the reader the feel of the subject. Throughout we think of probability as a proportion.

Example 1. In a complete natural population of poppies, I find that $\frac{1}{10}$ of the plants have five capsules. The probability that a plant in that population has five capsules is therefore $\frac{1}{10}$.

Example 2. In a particular backcross Aa × aa (in which all progeny genotypes are equally viable), Mendelian theory tells us that $\frac{1}{2}$ of the progeny will have the aa phenotype (note that the progeny is of unlimited size, since the cross may be repeated as many times as we like). The probability that a particular offspring of this cross will have the aa phenotype is therefore $\frac{1}{2}$.

In both examples we found the probability that an *individual* plant or animal had a specified characteristic. However, we can just as well find the probability that a group of individuals has some stated characteristic.

Example 3. We take every family of size four in the country and ask how many members of the family watch a particular television programme. The group is the family, the characteristic is the number of members of the family who watch the programme. If we find that the number of viewers is three in $\frac{1}{5}$ of the families, we say that the probability that a (randomly chosen) family has three viewers is $\frac{1}{5}$.

In this example, the number of groups (families), though large, had a definite value. However, we can use the same terminology when the number of groups is indefinitely large.

Example 4. We consider all the mouse sibships of size four that might ever be bred and ask: in what proportion of sibships would there be just one male? The group is the sibship, the characteristic the number of males. Our question then may be rephrased: what is the probability of getting just one male in a mouse sibship of size four?

Mean and variance

We need to define mean and variance with some care. If the reader has seen definitions that look rather different from those we shall give, he may rest assured that our definitions are just more careful versions of those with which he is familiar.

Suppose we have some characteristic (e.g. capsule number on a plant, number of males in a sibship size four) that takes discrete values (e.g. the number of males could be 0, 1, 2, 3, or 4). If we multiply every value by the probability of obtaining that value and add, we obtain the mean, average or expected value of the characteristic (the terms *mean*, *average*, *expected* are used interchangeably). The symbol E (for "expected") is often used here. If we write $Pr(x)$ for the probability of obtaining a particular value x of the characteristic, our mean may be written Ex and defined in symbols as follows:

$$Ex = \sum_x xPr(x)$$

where, as indicated, the sum is taken over all possible values of x.

The variance of the characteristic, written $V(x)$, is defined as

$$V(x) = E[x - Ex]^2$$

that is, the variance is the mean of the squares of deviations from the mean (or "mean square deviation" for short). It may be shown by simple algebra that our formula is equivalent to

$$V(x) = Ex^2 - (Ex)^2$$

A useful rule is as follows. Suppose we have found the mean Ex and variance $V(x)$ of some characteristic in some particular case. Now suppose we obtain a new characteristic by multiplying every x by the same constant c. We wish to calculate the mean $E(cx)$ and variance

$V(cx)$ for the new characteristic. It turns out that

$$E(cx) = cE(x), \quad V(cx) = c^2 V(x)$$

that is, just multiply the original mean and variance by c and c^2 respectively.

Note finally that if c is a constant, $Ec = c$ and $V(c) = 0$.

Probability distributions

The complete set of $Pr(x)$ for all possible x is called the *probability distribution of the characteristic*. It may be given as a table of figures, as a diagram, or as a mathematical formula from which the $Pr(x)$ may be calculated.

Binomial distribution

Suppose we have a group size k. Let A be some attribute. Any member of the group has probability p of having the attribute, probability q $(=1-p)$ of not having the attribute. The characteristic of the group we shall consider is the number of members having the attribute.

Now suppose we have an indefinitely large number of groups. Suppose every group is size k and that any member of any group has the same probability p of having attribute A. Then the probability that, in any group, r individuals have attribute A is

$$\frac{k!}{r!(k-r)!} q^{k-r} p^r$$

This is known as the binomial distribution.

Example 5. In chapter 3, we drew samples, size $2N$, of alleles from an infinite population of possible alleles with allele frequencies $\underline{A}\, p_0, \underline{a}\, q_0$. The members of any sample are alleles; if an allele is $\underline{A}$ we shall say it possesses attribute $\underline{A}$. Thus $k = 2N$, $p = p_0$, $q = q_0$ and the probability that a sample contains $r\, \underline{A}$ alleles is

$$\frac{(2N)!}{r!(2N-r)!} q_0^{2N-r} p_0^r$$

Now generally r will vary from one group to another, so that we may speak of the mean and variance of r. For the binomial distribution, these turn out to equal kp and kpq respectively. Thus in our example the mean number of $\underline{A}$ alleles is $2Np_0$, the variance of the number of $\underline{A}$ alleles is $2Np_0q_0$.

Poisson distribution

Imagine we have a space in which isolated events occur.

Example 6. The space is a Petri dish (containing complete medium) left open for, say, five minutes; an isolated event is the arrival of a fungal contaminant.

Imagine the space subdivided into a very large number of very small regions of equal area; suppose an event is equally likely to occur in any such region (this implies that events happen "independently at random").

Now imagine an indefinitely large number of such spaces having identical properties (in our example all Petri dishes would be the same size, the lids would be left off for the same time, and dishes would all be exposed to the same atmosphere). Let m be the mean number of events per space (dish). Then the probability that, in any given space, r events occur is

$$e^{-m}\frac{m^r}{r!}$$

(where e is the usual constant $2.71828\ldots$). This is the Poisson distribution.

Example 7. In chapter 2, the events are heterozygous tracts which appear in the space of all (conceptually) paired chromosomes in any pre-meiotic cell in a given line. The Poisson distribution will not apply until a fairly late stage in the inbreeding process; at this late stage heterozygous tracts are propagating independently.

Poisson approximation to binomial distribution

Consider a binomial distribution in which p is small, k large and kp moderate. Then the probability that, in any group, r individuals have attribute $\underline{A}$ can be closely approximated by

$$e^{-m}\frac{m^r}{r!}$$

where $m = kp$. For example, if $p = \frac{1}{200}$, $k = 200$ so that $kp = 1$ we find

r	Probability calculated from binomial	Probability calculated from Poisson
0	0.3670	0.3679
1	0.3688	0.3679
2	0.1844	0.1839
3	0.0612	0.0613
4	0.0151	0.0153
>4	0.0035	0.0037

Example 8. In chapter 3, we used this approximation in the case $p = 1/(2N)$, $k = 2N$. Obviously our approximation is very accurate when N exceeds 100.

Continuous distributions

So far we have considered cases where the characteristic took discrete values. If, however, the characteristic were, say, plant diameter or wing length in *Drosophila*, it could take any value within some range. Notice that, as in the cases discussed earlier in this appendix, the characteristic could be the property of an individual, such as the diameter of an individual plant, or the property of a group, such as the *mean* diameter of all plants derived from a specific cross.

A characteristic that can take any value within a given range is said to vary continuously. In such cases, awkward difficulties arise if we attempt to define the probability that the characteristic takes some definite value. We can, however, define without difficulty the probability that the characteristic lies in some range of values, say a to b. This latter probability can be represented as the area under a curve. The mean and variance of the characteristic can also be defined (save in rather bizarre circumstances) in a way that accords with intuitive understanding of the concept of mean and variance; we need not give details here.

Normal distribution

Suppose we have a continuously varying characteristic, with mean μ and variance σ^2. It often turns out, at least approximately, that the probability that the characteristic lies in the range of values a to b is the following area under a curve

$$\int_a^b \frac{1}{\sigma\sqrt{2\pi}} e^{-(x-\mu)^2/(2\sigma^2)} dx$$

where π is the usual $3.14159\ldots$, e is the usual $2.71828\ldots$ and $\sigma = \sqrt{\sigma^2}$. If σ^2 is small, almost all values of the characteristic will be close to the mean μ.

If our probability is given exactly by the formula just quoted, the characteristic is said to be *normally distributed*. If in such cases we obtain a new characteristic by multiplying all values of the old characteristic by the same constant c, the new characteristic is also normally distributed.

Normal approximation to binomial distribution

Consider a binomial distribution in which p is not too extreme and k is large. Then the probability that, in any group, the number of individuals having attribute $\underline{A}$ lies between a and b may be approximated by the formula for the normal distribution

$$\int_a^b \frac{1}{\sigma\sqrt{2\pi}} e^{-(x-\mu)^2/(2\sigma^2)} dx$$

provided we write $\mu = kp$, $\sigma^2 = kpq$ (these being the mean and variance of the binomially distributed characteristic). We used this approximation in chapter 3.

A note on chapter 3

We sketch the proof of a point which we merely quoted in chapter 3. Suppose that the frequency of allele $\underline{A}$ in a population in generation t is p_t and that frequencies are changing by drift only. If we used the argument given in chapter 3 for the case $t = 0$, we should find that in generation $(t+1)$

$$\text{mean allele frequency} = p_t$$
$$\text{mean frequency of heterozygotes} = 2p_t q_t \left(1 - \frac{1}{2N}\right)$$

After generation 0, however, p_t varies with population. By a standard statistical result, we allow for this by taking the mean, over all populations, of the right-hand sides of the above equations to give us the following correct results:

(a) mean allele frequency in $(t+1)$ = mean allele frequency in t (for any t); thus the mean allele frequency never changes and is always p_0.

(s) mean frequency of heterozygotes in $(t+1) = \{$mean frequency of heterozygotes in $t\} \times (1 - 1/(2N))$ (for any t); thus the mean frequency of heterozygotes declines by $(1 - 1/(2N))$ every generation and is therefore $2p_0 q_0 (1 - 1/(2N))^t$ in generation t.

Some mathematical points

Let e have its usual meaning. Then for any value of x

$$e^x = 1 + x + \frac{x^2}{2!} + \frac{x^3}{3!} + \frac{x^4}{4!} + \dots$$

in the sense that as we add more and more terms to the right-hand side, we get nearer and nearer to e^x. Similarly, when x lies between -1 and 1,

$$(1+x)^{-1} = 1 - x + x^2 - x^3 + x^4 - \dots$$
$$(1+x)^{-2} = 1 - 2x + 3x^2 - 4x^3 + 5x^4 - \dots$$
$$\log_e(1+x) = x - \tfrac{1}{2}x^2 + \tfrac{1}{3}x^3 - \tfrac{1}{4}x^4 + \dots$$

(the last result also holds when $x = 1$). If x is small, we can ignore terms in $x^2, x^3, x^4, \dots$ and our formulae provide useful simple approximations.

Put $x = -1/n$. When n is large, x is small so that $\log_e(1+x)$ is close to x and we have the close approximation

$$n \log_e \left(1 - \frac{1}{n} \right) = n \left(-\frac{1}{n} \right) = -1$$

and on taking antilogs we find

$$\left(1 - \frac{1}{n} \right)^n \text{ is close to } e^{-1}$$

when n is large. Similarly, we show that

$$\left(1 - \frac{1}{2N} \right)^t \text{ is close to } e^{-t/(2N)}$$

when N is large.

The formulae we have given for e^x, $(1+x)^{-1}$, etc., are, in fact, all special cases of a general formula, known as the Taylor series. Suppose y is a function of a variable x. When x takes a particular value a, write $f(a)$ for the value of y, $f'(a)$ for the value of dy/dx, $f''(a)$ for the value of d^2y/dx^2

and so on. Under a fairly wide range of circumstances, $f(a+h)$, the value of y when x takes another particular value $a+h$, is

$$f(a)+hf'(a) + \frac{h^2}{2!} f''(a) + \frac{h^3}{3!} f'''(a)+ \ldots$$

(to obtain our special cases, put a equal to 0 and h equal to x).

SUGGESTIONS FOR FURTHER READING

Section A. Particularly recommended (note, however, that these authors sometimes disagree with one another!)

Ayala, F. J. (ed.) (1976) *Molecular Evolution*. Sinauer Associates, Sunderland, Massachusetts.
Crow, J. F. and Kimura, M. (1970) *An Introduction to Population Genetics Theory*. Harper and Row, New York.
Dobzhansky, Th. (1970) *Genetics of the Evolutionary Process*. Columbia University Press, New York.
Ford, E. B. (1975) *Ecological Genetics*. 4th ed. Chapman and Hall, London.
Lewontin, R. C. (1974) *The Genetic Basis of Evolutionary Change*. Columbia University Press, New York.
Sheppard, P. M. (1975) *Natural Selection and Heredity*. 4th ed. Hutchinson, London.

Section B. References cited in the text.

Allard, R. W., Jain, S. K. and Workman, P. L. (1968) The genetics of inbreeding populations. *Advan. Genet.*, **14**, 55–131.
Allard, R. W. and Kahler, A. L. (1971) Allozyme polymorphisms in plant populations. *Stadler Symp.* (University of Missouri), **3**, 9–24.
Allard, R. W., Kahler, A. L. and Clegg, M. T. (1975) Isozymes in plant population genetics. In *Isozymes. IV. Genetics and Evolution* (ed. Markert, C. L.), Academic Press, New York, 261–272.
Allison, A. C. (1955) Aspects of polymorphism in Man. *Cold Spring Harbor Symp. Quant. Biol.*, **20**, 239–255.
Allison, A. C. (1964) Polymorphism and natural selection in human populations. *Cold Spring Harbor Symp. Quant. Biol.*, **29**, 137–149.
Anderson, W. W., Dobzhansky, Th., Pavlovsky, O., Powell, J. R. and Yardley, D. G. (1975) Genetics of natural populations. XLII. Three decades of genetic change in *Drosophila pseudoobscura. Evolution*, **29**, 24–36.
Arnold, R. W. (1971) *Cepaea nemoralis* on the East Sussex South Downs and the nature of area effects. *Heredity*, **26**, 277–298.
Beckenbach, A. T. and Prakash, S. (1977) Examination of allelic variation at the hexokinase loci of *Drosophila pseudoobscura* and *D. persimilis* by different methods. *Genetics*, **87**, 743–761.
Bennett, J. H. (1953) Junctions in inbreeding. *Genetica*, **26**, 392–406.
Bennett, J. H. (1954) The distribution of heterogeneity upon inbreeding. *J. Roy. Stat. Soc.*, *B*, **16**, 88–99.

Billingham, R. E., Brent, L., Medawar, P. B. and Sparrow, E. M. (1954) Quantitative studies on tissue transplantation immunity. I. The survival times of skin homografts exchanged between members of different inbred strains of mice. *Proc. Roy. Soc., B,* **143,** 43–58.

Bishop, J. A. (1972) An experimental study of the cline of industrial melanism in *Biston betularia* (L.) (Lepidoptera) between urban Liverpool and rural North Wales. *J. An. Ecol.,* **41,** 209–243.

Bishop, J. A. and Cook, L. M. (1975) Moths, melanism and clean air. *Sci. Am.,* **232,** 90–99.

Bodmer, W. F. (1965) Differential fertility in population genetics models. *Genetics,* **51,** 411–424.

Bodmer, W. F. and Bodmer, J. G. (1978) Evolution and function of the HLA system. *Br. Med. Bull.,* **34,** 309–316.

Boorman, S. A. and Levitt, P. R. (1973) Group selection on the boundary of a stable population. *Theor. Pop. Biol.,* **4,** 85–128.

Briscoe, D. A., Robertson, A. and Malpica, J. M. (1975) Dominance at Adh locus in response of adult *Drosophila melanogaster* to environmental alcohol. *Nature,* **255,** 148–149.

Cain, A. J. (1968) Sand-dune populations of *Cepaea nemoralis* (L.). *Phil. Trans. Roy. Soc., B,* **253,** 499–517.

Cain, A. J. (1971) Colour and banding morphs in subfossil samples of the snail *Cepaea.* In *Ecological Genetics and Evolution* (ed. Creed, R.). Blackwell, Oxford, 65–92.

Cain, A. J. and Currey, J. D. (1963) Area effects in *Cepaea. Phil. Trans. Roy. Soc., B,* **246,** 1–81.

Cain, A. J. and Currey, J. D. (1968) Ecogenetics of a population of *Cepaea nemoralis* subject to strong area effects. *Phil. Trans. Roy. Soc., B,* **253,** 447–482.

Cain, A. J. and Sheppard, P. M. (1950) Selection in the polymorphic land snail *Cepaea nemoralis* (L.). *Heredity,* **4,** 275–294.

Cain, A. J. and Sheppard, P. M. (1954) Natural selection in *Cepaea. Genetics,* **39,** 89–116.

Cain, A. J., Sheppard, P. M. and King, J. M. B. (1968) The genetics of some morphs and varieties of *Cepaea nemoralis* (L.). *Phil. Trans. Roy. Soc., B,* **253,** 383–396.

Carr, R. N. and Nassar, R. F. (1970) Effects of selection and drift on the dynamics of finite populations. I. Ultimate probability of fixation of a favorable allele. *Biometrics,* **26,** 41–49.

Caten, C. E. (1979) Quantitative genetic variation in fungi. In *Quantitative Genetic Variation* (ed. Thompson, J. N. and Thoday, J. M.). Academic Press. New York. 35–59.

Chia, A. B. (1968) Random mating in a population of cyclic size. *J. Appl. Prob.,* **5,** 21–30.

Clarke, B., Arthur, W., Horsley, D. T. and Parkin, D. T. (1978) Genetic variation and natural selection in pulmonate molluscs. In *Pulmonates,* 2A, (ed. Fretter, V. and Peake, J.), Academic Press, London, 219–270.

Clarke, B., Diver, C. and Murray, J. (1968) The spatial and temporal distribution of phenotypes in a colony of *Cepaea nemoralis* (L.). *Phil. Trans. Roy. Soc., B,* **253,** 519–548.

Clarke, B. and O'Donald, P. (1964) Frequency-dependent selection. *Heredity,* **19,** 201–206.

Clarke, C. A. and Sheppard, P. M. (1960) The evolution of dominance under disruptive selection. *Heredity,* **14,** 73–87.

Clarke, C. A. and Sheppard, P. M. (1971) Further studies on the genetics of the mimetic butterfly *Papilio memnon* L. *Phil. Trans. Roy. Soc., B,* **263,** 35–70.

Clarke, C. A. and Sheppard, P. M. (1972) The genetics of the mimetic butterfly *Papilio polytes* L. *Phil. Trans. Roy. Soc., B,* **263,** 431–458.

Coyne, J. A. and Felton, A. A. (1977) Genetic heterogeneity at two alcohol dehydrogenase loci in *Drosophila pseudoobscura* and *Drosophila.persimilis. Genetics,* **87,** 285–304.

Croft, J. H. and Jinks, J. L. (1977) Aspects of the population genetics of *Aspergillus nidulans.*

In *Genetics and Physiology of Aspergillus* (ed. Smith, J. E. and Pateman, J. A.), Academic Press, London, 339–360.

Crow, J. F. and Kimura, M. (1970) See Section A.

Crow, J. F. and Morton, N. E. (1955) Measurement of gene frequency drift in small populations. *Evolution*, **9**, 202–214.

Darwin, C. (1876) *The Effects of Cross and Self Fertilisation in the Vegetable Kingdom.* John Murray, London.

Davies, R. W. (1971) The genetic relationship of two quantitative characters in *Drosophila melanogaster*. II. Location of the effects. *Genetics*, **69**, 363–375.

Dawkins, R. (1976) *The Selfish Gene.* Oxford University Press, Oxford.

Day, T. H., Hillier, P. C. and Clarke, B. (1974) The relative quantities and catalytic activities of enzymes produced by alleles at the alcohol dehydrogenase locus in *Drosophila melanogaster*. *Bioch. Genet.*, **11**, 155–165.

Delden, W. van, Boerema, A. C. and Kamping, A. (1978) The alcohol dehydrogenase polymorphism in populations of *Drosophila melanogaster*. I. Selection in different environments. *Genetics*, **90**, 161–191.

Deol, M. S. Grüneberg, H., Searle, A. G. and Truslove, G. M. (1960) How pure are our inbred strains of mice? *Genet. Res.*, **1**, 50–58.

Dobzhansky, Th. (1970) See Section A.

Dobzhansky, Th. (1971) Evolutionary oscillations in *Drosophila pseudoobscura*. In *Ecological Genetics and Evolution*, (ed. Creed, R.), Blackwell, Oxford, 109–133.

Dowdeswell, W. H. (1975) *The Mechanism of Evolution* (4th ed.). Heinemann, London.

Dowdeswell, W. H. (1978) Darwinism and indoctrination. *Sch. Sci. Rev.*, **59**, 763–764.

East, E. M. and Jones, D. F. (1919) *Inbreeding and Outbreeding.* Lippincott, Philadelphia.

Edwards, A. W. F. (1977) *Foundations of Mathematical Genetics.* Cambridge University Press, Cambridge.

Ellis, W. M., Keymer, R. J. and Jones, D. A. (1976) On the polymorphism of cyanogenesis in *Lotus corniculatus* L. VI. Ecological studies in the Netherlands. *Heredity*, **36**, 245–251.

Ellis, W. M., Keymer, R. J. and Jones, D. A. (1977) On the polymorphism of cyanogenesis in *Lotus corniculatus* L. VIII. Ecological studies in Anglesey. *Heredity*, **39**, 45–65.

Ewens,, W. J. (1963) Numerical results and diffusion approximations in a genetic process. *Biometrika*, **50**, 241–249.

Ewens, W. J. (1964) The pseudo-transient distribution and its uses in genetics. *J. Appl. Prob.*, **1**, 141–156.

Ewens, W. J. (1967) Random sampling and the rate of gene replacement. *Evolution*, **21**, 657–663.

Ewens, W. J. (1967a) The probability of survival of a new mutant in a fluctuating environment. *Heredity*, **22**, 438–443.

Ewens, W. J. (1969) *Population Genetics.* Methuen, London.

Ewens, W. J. (1973) Conditional diffusion processes in population genetics. *Theor. Pop. Biol.*, **4**, 21–30.

Ewens, W. J. (1977) Selection and neutrality. In *Measuring Selection in Natural Populations* (ed. Christiansen, F. B. and Fenchel, T. M.), Springer-Verlag, Berlin, 159–175.

Ewens, W. J. and Feldman, M. W. (1976) The theoretical assessment of selective neutrality. In *Population Genetics and Ecology* (ed. Karlin, S. and Nevo, E.), Academic Press, New York, 303–337.

Falconer, D. S. (1964) *Introduction to Quantitative Genetics.* Oliver and Boyd, Edinburgh.

Felsenstein, J. (1976) The theoretical population genetics of variable selection and migration. *Ann. Rev. Genet.*, **10**, 253–280.

Fisher, R. A. (1922) On the dominance ratio. *Proc. Roy. Soc. Edinb.*, **42**, 321–341.

Fisher, R. A. (1928) The possible modification of the response of the wild type to recurrent mutations. *Amer. Nat.*, **62**, 115–126.

Fisher, R. A. (1930) *The Genetical Theory of Natural Selection.* 1st. ed., Clarendon Press, Oxford. 2nd. ed. (1958), Dover Publications, New York.

Fisher, R. A. (1930*a*) The distribution of gene ratios for rare mutations. *Proc. Roy. Soc. Edinb.*, **50**, 205–220.

Fisher, R. A. (1949, 1965) *The Theory of Inbreeding.* Oliver and Boyd, Edinburgh.

Fisher, R. A. (1954) A fuller theory of "junctions" in inbreeding. *Heredity*, **8**, 187–197.

Fisher, R. A. (1959) An algebraically exact examination of junction formation and transmission in parent-offspring inbreeding. *Heredity*, **13**, 179–186.

Fisher, R. A. and Ford, E. B. (1947) The spread of a gene in natural conditions in a colony of the moth *Panaxia dominula* L. *Heredity*, **1**, 143–174.

Fisher, R. A. and Holt, S. B. (1944) The experimental modification of dominance in Danforth's short-tailed mutant mice. *Ann. Eugen.*, **12**, 102–120.

Fitch, W. M. (1976) Molecular evolutionary clocks. In *Molecular Evolution* (see Section A), 160–178.

Fitch, W. M. and Langley, C. H. (1976) Protein evolution and the molecular clock. *Fed. Proc.*, **35**, 2092–2097.

Ford, E. B. (1940) Genetic research in the Lepidoptera. *Ann. Eugen.*, **10**, 227–252.

Ford, E. B. (1955) Polymorphism and taxonomy. *Heredity*, **9**, 255–264.

Ford, E. B. (1975) See Section A.

Ford, E. B. and Sheppard, P. M. (1969) The medionigra polymorphism of *Panaxia dominula. Heredity*, **24**, 561–569.

Franklin, I. R. (1977) The distribution of the proportion of the genome which is homozygous by descent in inbred individuals. *Theor. Pop. Biol.*, **11**, 60–80.

Gale, J. S., Rana, M. S. and Lawrence, M. J. (1974) Variation in wild populations of *Papaver dubium*. IX. Limited possibilities for assortative mating. *Heredity*, **32**, 389–396.

Gale, J. S., Solomon, R., Thomas, W. T. B. and Zuberi, M. I. (1976) Variation in wild populations of *Papaver dubium*. XI. Further studies on direction of dominance. *Heredity*, **36**, 417–422.

Gibson, J. (1970) Enzyme flexibility in *Drosophila melanogaster. Nature*, **227**, 959–960.

Gillespie, J. H. (1977) A general model to account for enzyme variation in natural populations. III. Multiple alleles. *Evolution*, **31**, 85–90.

Gillespie, J. H. and Langley, C. H. (1974) A general model to account for enzyme variation in natural populations. *Genetics*, **76**, 837–848.

Goodhart, C. B. (1963) "Area effects" and non-adaptive variation between populations of *Cepaea* (Mollusca). *Heredity*, **18**, 459–465.

Haigh, J. and Maynard Smith, J. (1972) Population size and protein variation in man. *Genet. Res.*, **19**, 73–89.

Haldane, J. B. S. (1924) A mathematical theory of natural and artificial selection. Part I. *Trans. Camb. Phil. Soc.*, **23**, 19–41.

Haldane, J. B. S. (1924*a*) A mathematical theory of natural and artificial selection. II. The influence of partial self-fertilisation, inbreeding, assortative mating, and selective fertilisation on the composition of Mendelian populations, and on natural selection. *Proc. Camb. Phil. Soc.*, (*Biol. Sci.*) (later *Biological Reviews*), **1**, 158–163.

Haldane, J. B. S. (1927) A mathematical theory of natural and artificial selection. V. Selection and mutation. *Proc. Camb. Phil. Soc.*, **23**, 838–844.

Haldane, J. B. S. (1932) A mathematical theory of natural and artificial selection. IX. Rapid selection. *Proc. Camb. Phil. Soc.*, **28**, 244–248.

Haldane, J. B. S. (1932*a*) *The Causes of Evolution.* Longmans Green, London.

Haldane, J. B. S. (1936) The amount of heterozygosis to be expected in an approximately pure line. *J. Genet.*, **32**, 375–391.

Haldane, J. B. S. (1937) Some theoretical results of continued brother–sister mating. *J. Genet.*, **34**, 265–274.

Haldane, J. B. S. (1939) The theory of evolution of dominance. *J. Genet.*, **37**, 365–374.

Haldane, J. B. S. (1940) The conflict between selection and mutation of harmful recessive genes. *Ann. Eugen.*, **10**, 417–421.

Haldane, J. B. S. (1955) Population genetics. *New Biology*, **18**, 34–51.

Haldane, J. B. S. (1956) The conflict between inbreeding and selection. I. Self-fertilisation. *J. Genet.*, **54**, 56–63.

Haldane, J. B. S. and Jayakar, S. D. (1963) The solution of some equations occurring in population genetics. *J. Genet.*, **58**, 291–317.

Hamilton, W. D. (1964) The genetical evolution of social behaviour. I and II. *J. Theoret. Biol.*, **7**, 1–52.

Harris, H. (1966) Enzyme polymorphisms in Man. *Proc. Roy. Soc.*, B, **164**, 298–310.

Harris, H. (1975) *The Principles of Human Biochemical Genetics.* 2nd ed. North-Holland Publishing Company, Amsterdam.

Harris, H. and Hopkinson, D. A. (1978) *Handbook of Enzyme Electrophoresis in Human Genetics.* North-Holland Publishing Company, Amsterdam.

Harvey, P. H. (1976) Factors influencing the shell pattern polymorphism of *Cepaea nemoralis* (L.) in East Yorkshire: a test case. *Heredity*, **36**, 1–10.

Hayman, B. I. (1953) Mixed selfing and random mating when homozygotes are at a disadvantage. *Heredity*, **7**, 185–192.

Hayman, B. I. and Mather, K. (1953) The progress of inbreeding when homozygotes are at a disadvantage. *Heredity*, **7**, 165–183.

Hayman, B. I. and Mather, K. (1956) Inbreeding when homozygotes are at a disadvantage: a reply. *Heredity*, **10**, 271–274.

Humphreys, M. O. and Gale, J. S. (1974) Variation in wild populations of *Papaver dubium*. VIII. The mating system. *Heredity*, **33**, 33–41.

John, B. and Lewis, K. R. (1965) *The Meiotic System* (*Protoplasmatologia*, VI, F 1). Springer Verlag, Vienna.

Jones, D. A. (1972) On the polymorphism of cyanogenesis in *Lotus corniculatus* L. IV. The Netherlands. *Genetica*, **43**, 394–406.

Jones, J. S. (1973) Ecological genetics and natural selection in molluscs. *Science*, **182**, 546–552.

Karlin, S. (1975) General two-locus selection models: some objectives, results and interpretations. *Theor. Pop. Biol.*, **7**, 364–398.

Karlin, S. and Carmelli, D. (1975) Numerical studies on two-loci selection models with general viabilities. *Theor. Pop. Biol.*, **7**, 399–421.

Kearsey, M. J. and Barnes, B. W. (1970) Variation for metrical characters in Drosophila populations. II. Natural selection. *Heredity*, **25**, 11–21.

Kearsey, M. J. and Kojima, K. I. (1967) The genetic architecture of body weight and egg hatchability in *Drosophila melanogaster*. *Genetics*, **56**, 23–37.

Kempthorne, O. (1957) *An Introduction to Genetic Statistics.* Wiley, New York.

Kettlewell, H. B. D. (1965) Insect survival and selection for pattern. *Science*, **148**, 1290–1296.

Kettlewell, H. B. D. (1973) *The Evolution of Melanism.* Clarendon Press, Oxford.

Kimura, M. (1955) Solution of a process of random genetic drift with a continuous model. *Proc. Nat. Acad. Sci.*, **41**, 144–150.

Kimura, M. (1956) Rules for testing stability of a selective polymorphism. *Proc. Nat. Acad. Sci.*, **42**, 336–340.

Kimura, M. (1957) Some problems of stochastic processes in genetics. *Ann. Math. Stat.*, **28**, 882–901.

Kimura, M. (1964) Diffusion models in population genetics. *J. Appl. Prob.*, **1**, 177–232.

Kimura, M. (1968) Evolutionary rate at the molecular level. *Nature*, **217**, 624–626.

Kimura, M. (1968a) Haldane's contributions to the mathematical theories of evolution and population genetics. In *Haldane and Modern Biology* (ed. Dronamraju, K. R.), The Johns Hopkins Press, Baltimore, 133–140.

Kimura, M. (1969) The rate of molecular evolution considered from the standpoint of population genetics. *Proc. Nat. Acad. Sci.*, **63**, 1181–1188.

Kimura, M. and Crow, J. F. (1964) The number of alleles that can be maintained in a finite population. *Genetics*, **49**, 725–738.

Kimura, M. and Maruyama, T. (1971) Pattern of neutral polymorphism in a geographically structured population. *Genet. Res.*, **18**, 125–131.

Kimura, M. and Ohta, T. (1969) The average number of generations until fixation of a mutant gene in a finite population. *Genetics*, **61**, 763–771.

Kimura, M. and Ohta, T. (1971) Protein polymorphism as a phase of molecular evolution. *Nature*, **229**, 467–469.

Kojima, K. I. and Kelleher, T. M. (1962) Survival of mutant genes. *Amer. Nat.*, **96**, 329–346.

Lakatos, I. (1976) *Proofs and Refutations*. Cambridge University Press, Cambridge.

Langley, C. H. and Fitch, W. M. (1974) An examination of the constancy of the rate of molecular evolution. *J. Mol. Evol.*, **3**, 161–177.

Lees, D. R. and Creed, E. R. (1975) Industrial melanism in *Biston betularia*: the role of selective predation. *J. An. Ecol.*, **44**, 67–83.

Levin, D. A. (1978) Some genetic consequences of being a plant. In *Ecological Genetics: The Interface* (ed. Brussard, P. F.), Springer-Verlag, New York, 189–212.

Levins, R. (1970) Extinction. In *Some Mathematical Questions in Biology* (ed. Gerstenhaber, M.) American Mathematical Society, Providence, 77–107.

Lewontin, R. C. (1974) See Section A.

Lewontin, R. C. and Cockerham, C. C. (1959) The goodness-of-fit test for detecting natural selection in random mating populations. *Evolution*, **13**, 561–564.

Lewontin, R. C., Ginzburg, L. R. and Tuljapurkar, S. D. (1978) Heterosis as an explanation for large amounts of genic polymorphism. *Genetics*, **88**, 149–170.

Lewontin, R. C. and Hubby, J. L. (1966) A molecular approach to the study of genic heterozygosity in natural populations. II. Amount of variation and degree of heterozygosity in natural populations of *Drosophila pseudoobscura*. *Genetics*, **54**, 595–609.

Linney, R., Barnes, B. W. and Kearsey, M. J. (1971) Variation for metrical characters in Drosophila populations. III. The nature of selection. *Heredity*, **27**, 163–174.

McDonald, J. F. and Ayala, F. J. (1974) Genetic response to environmental heterogeneity. *Nature*, **250**, 572–574.

McKenzie, J. A. and McKechnie, S. W. (1978) Ethanol tolerance and the Adh polymorphism in a natural population of *Drosophila melanogaster*. *Nature*, **272**, 75–76.

McKusick, V. A. (1974) Genetic studies in American inbred populations with particular reference to the Old Order Amish. In *Genetic Polymorphisms and Diseases in Man* (ed. Ramot, B.), Academic Press, New York, 150–158.

McKusick, V. A., Hostetler, J. A., Egeland, J. A. and Eldridge, R. (1964) The distribution of certain genes in the Old Order Amish. *Cold Spring Harbor Symp. Quant. Biol.*, **29**, 99–114.

Malécot, G. (1948) *Les Mathématiques de l'Hérédité*. Translation (*The Mathematics of Heredity*) by Yermanos, D. M. (1969). Freeman and Company, San Francisco.

Mandel, S. P. H. (1959) The stability of a multiple allelic system. *Heredity*, **13**, 289–302.

Marshall, D. R. and Allard, R. W. (1970) Isozyme polymorphisms in natural populations of *Avena fatua* and *A. barbata*. *Heredity*, **25**, 373–382.

Maruyama, T. (1970) On the rate of decrease of heterozygosity in circular stepping stone models of populations. *Theor. Pop. Biol.*, **1**, 101–119.

Maruyama, T. (1970*a*) Effective number of alleles in a subdivided population. *Theor. Pop. Biol.*, **1**, 273–306.

Maruyama, T. (1972) The average number and the variance of generations at particular gene frequency in the course of fixation of a mutant gene in a finite population. *Genet. Res.*, **19**, 109–113.

Maruyama, T. (1977) *Stochastic Problems in Population Genetics*. Springer-Verlag, Berlin.

Mather, K. (1960) Evolution in polygenic systems. *Int. Colloquium on Evolution and Genetics*, Acad. Naz. dei Lincei, Rome, 131–152.

Mather, K. (1973) *Genetical Structure of Populations*. Chapman and Hall, London.

Maynard Smith, J. (1964) Group selection and kin selection: a rejoinder. *Nature*, **201**, 1145–1147.

Maynard Smith, J. (1969) The status of neo-Darwinism. In *Towards a Theoretical Biology*, 2, (ed. Waddington, C. H.), Edinburgh University Press, 82–89.

Maynard Smith, J. (1970) Population size, polymorphism and the rate of non-darwinian evolution. *Amer. Nat.*, **104**, 231–237.

Maynard Smith, J. (1977) The limitations of evolutionary theory. In *The Encyclopaedia of Ignorance* (ed. Duncan, R. and Weston-Smith, M.), Pergamon Press, Oxford, 235–242.

Maynard Smith, J. (1978) *The Evolution of Sex*. Cambridge University Press, Cambridge.

Maynard Smith, J. and Haigh, J. (1974) The hitch-hiking effect of a favourable gene. *Genet. Res.*, **23**, 23–35.

Mayr, E. (1963) *Animal Species and Evolution*. Harvard University Press.

Mendel, G. (1865) Experiments in plant hybridisation. English translation reproduced in *Experiments in Plant Hybridisation* (ed. Bennett, J. H.). Oliver and Boyd, Edinburgh, 7–51.

Milkman, R. (1975) Allozyme variation in *E. coli* of diverse natural origins. In *Isozymes. IV. Genetics and Evolution* (ed. Markert, C. L.), Academic Press, New York, 273–285.

Minawa, A. and Birley, A. J. (1978) The genetical response to natural selection by varied environments. I. Short-term observations. *Heredity*, **40**, 39–50.

Moorhead, P. S. and Kaplan, M. M. (eds.) (1967) *Mathematical Challenges to the Neo-Darwinian Interpretation of Evolution*. The Wistar Institute Press, Philadelphia.

Moran, P. A. P. (1960) The survival of a mutant gene under selection. II. *J. Australian Math. Soc.*, **1**, 485–491.

Mukai, T. and Cockerham, C. C. (1977) Spontaneous mutation rates at enzyme loci in *Drosophila melanogaster*. *Proc. Nat. Acad. Sci.*, **74**, 2514–2517.

Nei, M. (1975) *Molecular Population Genetics and Evolution*. North-Holland Publishing Company, Amsterdam.

Nei, M. and Chakraborty, R. (1976) Empirical relationship between the numbers of nucleotide substitutions and interspecific identity of amino acid sequences in some proteins. *J. Mol. Evol.*, **7**, 313–323.

Nei, M. and Li, W. H. (1975) Probability of identical monomorphism in related species. *Genet. Res.*, **26**, 31–43.

Noda, S. (1968) Achiasmate bivalent formation by parallel pairing in PMCs of *Fritillaria amabilis. Bot. Mag. Tokyo*, **81**, 344–345.

O'Brien, S. J. and MacIntyre, R. J. (1969) An analysis of gene-enzyme variability in natural populations of *Drosophila melanogaster* and *D. simulans. Amer. Nat.*, **103**, 97–113.

O'Donald, P. (1968) Natural selection by glow-worms in a population of *Cepaea nemoralis. Nature*, **217**, 194.

Ohta, T. (1974) Mutational pressure as the main cause of molecular evolution and polymorphism. *Nature*, **252**, 351–354.

Ohta, T. (1976) Role of very slightly deleterious mutations in molecular evolution and polymorphism. *Theor. Pop. Biol.*, **10**, 254–275.

Ohta, T. and Kimura, M. (1973) A model of mutation appropriate to estimate the number of electrophoretically detectable alleles in a finite population. *Genet. Res.*, **22**, 201–204.

Ohta, T. and Kimura, M. (1975) The effect of selected linked locus on heterozygosity of neutral alleles (the hitch-hiking effect). *Genet. Res.*, **25**, 313–326.

Parker, G. A. (1978) Selfish genes, evolutionary games, and the adaptiveness of behaviour. *Nature*, **274**, 849–855.

Powell, J. R. (1971) Genetic polymorphism in varied environments. *Science*, **174**, 1035–1036.

Prakash, S., Lewontin, R. C. and Hubby, J. L. (1969) A molecular approach to the study of genic heterozygosity in natural populations. IV. Patterns of genic variation in central, marginal and isolated populations of *Drosophila pseudoobscura*. *Genetics*, **61**, 841–858.

Prout, T. (1971) The relation between fitness components and population prediction in Drosophila. I. The estimation of fitness components. *Genetics*, **68**, 127–149.

Race, R. R. and Sanger, R. (1975) *Blood Groups in Man*. 6th ed. Blackwell, Oxford.

Rees, H. and Thompson, J. B. (1956) Genotypic control of chromosome behaviour in rye. III. Chiasma frequency in homozygotes and heterozygotes. *Heredity*, **10**, 409–424.

Reeve, E. C. R. (1955) Inbreeding with homozygotes at a disadvantage. *Ann. Hum. Genet.*, **19**, 332–346.

Reeve, E. C. R. (1957) Inbreeding with selection and linkage. I. Selfing. *Ann. Hum. Genet.*, **21**, 277–288.

Robertson, A. (1966) Artificial selection in plants and animals. *Proc. Roy. Soc.*, B, **164**, 341–349.

Scharloo, W., van Dijken, F. R., Hoorn, A. J. W., de Jong, G. and Thörig, G. E. W. (1977) Functional aspects of genetic variation. In *Measuring Selection in Natural Populations* (ed. Christiansen, F. B. and Fenchel, T. M.), Springer-Verlag, Berlin, 131–147.

Selander, R. K. (1976) Genic variation in natural populations. In *Molecular Evolution* (see Section A), 21–45.

Sing, C. F., Brewer, G. J. and Thirtle, B. (1973) Inherited biochemical variations in *Drosophila melanogaster*: noise or signal? I. Single locus analyses. *Genetics*, **75**, 381–404.

Singh, R. S., Lewontin, R. C. and Felton, A. A. (1976) Genetic heterogeneity within electrophoretic "alleles" of xanthine dehydrogenase in *Drosophila pseudoobscura*. *Genetics*, **84**, 609–629.

Soulé, M. (1976) Allozyme variation: its determinants in space and time. In *Molecular Evolution* (see Section A), 60–77.

Stebbins, G. L. (1971) Adaptive radiation of reproductive characteristics in Angiosperms. II. Seeds and seedlings. *Ann. Rev. Ecol. Sys.*, **2**, 237–260.

Stockmeyer, L. J. and Chandra, A. K. (1979) Intrinsically difficult problems. *Sci. Am.*, **240**, 124–133.

Stone, W. S., Wheeler, M. R., Johnson, F. M. and Kojima, K. I. (1968) Genetic variation in natural island populations of members of the *Drosophila nasuta* and *D. ananassae* subgroups. *Proc. Nat. Acad. Sci.*, **59**, 102–109.

Thomas, W. T. B. and Gale, J. S. (1977) Variation in wild populations of *Papaver dubium*. XII. Direction of dominance during development. *Heredity*, **39**, 305–312.

Thomson, G. (1977) The effect of a selected locus on linked neutral loci. *Genetics*, **85**, 753–788.

Trivers, R. L. and Hare, H. (1976) Haplodiploidy and the evolution of the social insects. *Science*, **191**, 249–263.

Turner, J. R. G. (1977) Butterfly mimicry: the genetical evolution of an adaptation. In *Evolutionary Biology*, **10**, (ed. Hecht, M. K., Steere, W. C., and Wallace, B.), Plenum Publishing Corporation, New York, 163–206.

Van Valen, L. (1974) Molecular evolution as predicted by natural selection. *J. Mol. Evol.*, **3**, 89–101.

Wallace, B. and Madden, C. (1953) The frequencies of sub- and supervitals in experimental populations of *Drosophila melanogaster*. *Genetics*, **38**, 456–470.

Williams, G. C. (1966) *Adaptation and Natural Selection*. Princeton University Press.

Wilson, A. C., Carlson, S. S. and White, T. J. (1977) Biochemical evolution. *Ann. Rev. Biochem.*, **46**, 573–639.

Wilson, E. O. (1975) *Sociobiology*. Harvard University Press.

Workman, P. L., Blumberg, B. S. and Cooper, A. J. (1963) Selection, gene migration and polymorphic stability in a U.S. White and Negro population. *Amer. J. Hum. Genet.*, **15**, 429–437.

Workman, P. L. and Jain, S. K. (1966) Zygotic selection under mixed random mating and self-fertilisation: Theory and problems of estimation. *Genetics*, **54**, 159–171.

Wright, S. (1931) Evolution in Mendelian populations. *Genetics*, **16**, 97–159.

Wright, S. (1937) The distribution of gene frequencies in populations. *Proc. Nat. Acad. Sci.*, **23**, 307–320.

Wright, S. (1939) Statistical genetics in relation to evolution. *Act. Scient. et Indus.*, **802**, 1–64.

Wright, S. (1948) On the roles of directed and random changes in gene frequency in the genetics of populations. *Evolution*, **2**, 279–294.

Wright, S. (1969, 1977, 1978) *Evolution and the Genetics of Populations* Vol. 2, 1969: Vol. 3, 1977: Vol. 4, 1978. University of Chicago Press.

Yamazaki, T. (1971) Measurement of fitness at the esterase-5 locus in *Drosophila pseudoobscura*. *Genetics*, **67**, 579–603.

Zuckerkandl, E. and Pauling, L. (1962) Molecular disease, evolution and genic heterogeneity. In *Horizons in Biochemistry* (ed. Kasha, M. and Pullman, B.), Academic Press, New York, 189–225.

Zuckerkandl, E. and Pauling, L. (1965) Evolutionary divergence and convergence in proteins. In *Evolving Genes and Proteins* (ed. Bryson, V. and Vogel, H. J.), Academic Press, New York, 97–166.

INDEX

185